DANGEROUS LESSONS

The Art and Science
of Teaching Military Tactics

DANGEROUS LESSONS

The Art and Science
of Teaching Military Tactics

by
Colonel (Ret'd) Charles S. Oliviero
&
Lieutenant-Colonel (Ret'd) Phil Halton

Library and Archives Canada Cataloguing in Publication

Oliviero, Charles S., and Halton, Phil, authors
Dangerous Lessons / Charles S. Oliviero and Phil Halton

Issued in print and electronic formats.

ISBN: 978-1-998501-55-7 (paperback)
ISBN: 978-1-998501-56-4 (ebook)

Cover Design: Pablo Javier Herrera
Interior Design: Winston S. Prescott

Double Dagger Books Ltd.
Toronto, Ontario, Canada
www.doubledagger.ca

DEDICATION

We dedicate this text to military instructors and their students everywhere.

TABLE OF CONTENTS

Part Three - Field Training

Part Four - Simulation

Part Five - Conclusions

FOREWORD

Forewords are normally written by subject matter experts, or somebody famous. I am neither. I am but a humble practitioner of tactics who has discovered over the course of almost four decades of military service there are three fundamental and inextricably linked pillars to what we call tactics that in my view must be understood if one aspires to succeed in this field.

First, tactics is not a single concept, but rather a process that results in a course of action ensuring mission success, whether on the plains of Afghanistan, the steppes of Ukraine, or the skies over Israel. This process involves combining three factors — a thorough knowledge of one's own capabilities as well as those of opposing forces, a clear understanding of the mission at hand, and a comprehensive grasp of the battle space — to decide what actions must be taken to achieve victory. Drills and standing operating procedures (SOPs) are vital enablers of this process, but they are not solutions on their own.

Second, tactics evolve. This evolution could be the result of previous failures, the introduction of new capabilities in the battle space (e.g., drones), or even new actions of opposing forces. Continual feedback is thus a vital component of not only the development of new enablers but of the entire aforementioned process itself.

Finally, tactics can be learned, and if they can be learned they can be taught, which is why training — constant, dynamic and imaginative — underpins all tactical successes at every level.

With this book Chuck Oliviero and Phil Halton provide an outstanding tool that can be used by new commanders to develop and hone their tactical skills, along with insightful examples of training for tactical success. It should be mandatory reading for all new commanders, regardless of rank. I only wish it had been available at the outset of my career rather than long after it ended.

Lieutenant-Colonel (Ret'd) Stephen Moffat, MSM, CD
Late of Lord Strathcona's Horse (Royal Canadians) and
The Royal Canadian Dragoons

FOR TRAINING INSTITUTIONS

We wrote this book to support military training institutions, schools, and unit-level instructors in the development and delivery of tactical education. It is not a doctrinal publication, nor does it seek to replace official training materials. Rather, it is a practical supplement — a guide to help instructors at all levels build understanding in their students and subordinates, especially in the teaching of small-unit tactics.

Dangerous Lessons can be used in whole or in part, and its modular structure allows for selective integration into classroom teaching, field instruction, or professional development programs. Each method outlined in this book includes a breakdown of what it teaches, how it teaches it, and how it can be applied with minimal resources. The inclusion of real-world examples, practical checklists, and reflection tools make it ideal for both formal and informal learning environments.

We encourage training institutions to make this text available to:

- Instructor development programs
- NCO and officer professional military education (PME) courses
- Platoon commander and section commander courses
- Field schools and combat training centers
- Pre-deployment and work-up training teams

If adopted formally, instructors may wish to:

- Assign specific chapters as pre-reading for instructional technique modules
- Use included exercises and examples as templates for practical lessons
- Encourage students to build and deliver their own training sessions using the methods outlined

The authors welcome the adaptation, extension, and critique of the ideas contained here. The battlespace is always changing, and so our tactical methods should as well. What matters most is that we keep teaching, keep learning, and continue to build military cultures where tactical thinking is both encouraged and understood — not just memorized.

For inquiries about institutional licensing, bulk orders, or supporting materials, contact the publisher at www.doubledagger.ca.

PREFACE

Training, teaching, educating, mentoring. These are all part of the many duties and responsibilities of a leader. Irrespective of what branch of service or what arm within that branch, it falls to the leader, whatever his or her rank, to perform all of these four functions — and more. These functions are not discrete, they are not independent of each other or in any way mutually exclusive. There is overlap, commonality, intersection, and reinforcement among them. But what is paramount for the leader to appreciate is that whatever function he, or she, is performing at any given time, the foundation of that leader's success is understanding.

Aristotle is famously misquoted regarding teaching. The misquote is: "Those who can, do; those who can't, teach." Nothing could be further from what the great philosopher, and mentor of Alexander the Great, actually said. It is so important that we'll let it stand alone.

Those who know, do; those who understand, teach.

The basis of your tactical success is understanding, but it is not just your understanding. It is also the understanding of your subordinates, or pupils. It falls to you to lead them to the goal of not just being able to do, but more importantly, to understand how they got there.

In this volume, we are opening the door to help you achieve that goal. Between the two of us, we have decades of command as well as instructional experience, and we have chosen to use that experience to guide those who wish to become better at their profession. Specifically, we will be focusing on how to teach tactics, but as already said, teaching is inextricably bound with all the other duties of leadership, and occasionally we'll comment on that as well.

There is much magical thinking about tactics. Certainly, history offers occasional tactical geniuses who seemed to have innately understood how to control troops in battle and achieve victory. But genius is exceedingly rare, and we need to consider if it was actually genius, or the result of a deeper understanding of the issues they faced, when compared with those around them. In this book we are offering a starting point for those individuals who not only want to better understand tactics, but also those who want to understand how to teach tactics.

Contrary to the old adage, tactics are not just the opinion of the senior commander present. Tactics, and their teaching, are the purview of all leaders at all levels. But before you can achieve the understanding that you'll need to teach tactics, you will need to put aside any magical thinking; you'll need to look behind the curtain. That is what we do here. We move the curtain aside so that we can dispel popular illusions.

German Field Marshal Erwin Rommel, the infamous commander of the Afrika Korps, once said, "Training is the best form of troop welfare; it prevents casualties." In this, the Field Marshal was correct. The more highly trained your soldiers are, the higher are their chances not only for success, but also for survival. But this training is not only what is demanded, or delivered, from above; it is also the training that you perform of your own volition. It all begins with you, the junior leader: applying a philosophy of leadership that is based on knowledge, the quest for excellence, and a deeper understanding of your profession.

Col. (Ret'd) CS Oliviero, CD, PhD   LCol.(Ret'd) PJ Halton, CD, MA

INTRODUCTION

Although not always seen in that light, military forces, whether on land, at sea, or in the air, are institutions of teaching and learning. When not engaged in combat, they are constantly training to do so, whether in the classroom, in simulators, or in their element on their equipment. Frederick the Great called his army "the school of the nation" where individuals would be educated and trained not only to fight but also to be better citizens.

None of this training, teaching, and learning happens by accident. It is considered, planned, and carefully executed. But there is a gap in this methodology, and this book aims to fill that gap. The gap we refer to is the teaching of tactics at the lower levels of military organizations. Non-commissioned officers (NCOs) and junior commissioned officers do the majority of this teaching. In spite of carrying this heavy burden, it is rarely the case that any of these leaders have been taught how to teach the tactics that they, and their subordinates, are expected to execute or pass on to others.

That is where this book comes in.

Before we begin, let's pause for a word on definitions. We have already said that training, teaching, educating, and mentoring are all related to each other. Strictly speaking, they can be quite different. Training a platoon to perform an ambush drill is quite different from *teaching* the platoon that same drill, educating them on why we follow certain strict rules during an ambush, or mentoring the individual soldiers on how better to perform their individual duties during the conduct of the ambush. But this book is not meant to be an academic guide. It is designed to assist junior leaders in their understanding of how best to teach tactics. The military uses the term "training" very broadly to encompass all aspects of the processes mentioned above, and so, with that in mind, we have chosen to most often use the term "training" in its broadest military sense. But remember, the goal will always be *understanding*.

Training does not need to be complicated, logistically expensive, or overly scripted. A good leader always has a short, sharp training plan in his or her back pocket, that can be used should an opportunity to teach a tactical lesson arise. Despite the fact that time is always the training resource in shortest supply, military scheduling often creates numerous pockets of "dead" time where troops are sitting and waiting. A savvy leader can capitalize on these periods of otherwise wasted time to conduct simple training that engages their subordinates. Often, this additional material reinforces the other training that is conducted at the direction of higher headquarters. By setting this example, you will also enable and encourage your subordinate leaders to do the same. Imagine if every leader, at all levels, used the dead time in the training calendar to good effect. The impact would be incredible, creating an exponential increase in the actual time spent training. This kind of training culture is a powerful force multiplier.

Naturally, if you are a newly promoted NCO or freshly commissioned second lieutenant, capitalizing on these impromptu teaching moments is a challenge. But with practice and some preparation, these moments will come to serve you (and your soldiers) well.

With all of this in mind, let's get started!

PART ONE
UNDERSTANDING TACTICS

MAKING SENSE OF TACTICS

In business, there is talk of marketing and operational "tactics." Clothing is tactical, gear is tactical, and even everyday items like backpacks and flashlights are labeled as tactical. But as military professionals, we should know what the word "tactics" actually means to us.

The United States Department of Defense defines tactics as:

> *1. The employment of units in combat.*
> *2. The ordered arrangement and maneuver of units in relation to each other and/or to the enemy in order to use their full potentialities.*

Neither of these statements shed much light on the matter, especially when considering tactics as practised at the company level and below. The "arrangement and maneuver" of units is clearly critical, but it does not explain how leaders should develop and apply these arrangements.[1] A better definition is needed for our purposes.

The United States Marine Corps defines tactics as:

> *...the art and science of winning engagements and battles. It includes the use of firepower and maneuver, the integration of different arms and the immediate exploitation of success to defeat the enemy...*

This definition feels much closer to what we need. There are some practical aspects listed, such as the *integration of different arms* and the *exploitation of*

---

[1] In professional military language, "maneuver" is understood to incorporate both firepower and movement.

success, not to mention the focus on winning *battles* and *defeating the enemy*. But how do you train to win battles? How do you determine how to use firepower and movement together, not to mention where and when? The suggestion that tactics are both an art and a science seems to pull the concept into the higher realms of human achievement, alongside music and painting. Does this help us with understanding what tactics are at the company level and below? Can we expect junior leaders to grasp this idea and just run with it?

British General Sir William Slim, renowned for his clear-sightedness and straight forward approach, describes it in a simpler and more accessible way:

> *There is only one principle of war and that's this. Hit the other fellow, as quick as you can, and as hard as you can, where it hurts him the most, when he ain't looking."*

This is a definition that every soldier can understand, and which begins to suggest how and where and when violence needs to be applied to achieve the desired effect. But even with this definition, we are no closer to understanding the practicality of how to actually train leaders and soldiers in the "art and science" of tactics.

Although we have not landed on the ideal definition, we now have some anchor points upon which to tie our ideas. Before we move on, allow us to offer one last definition from the book *Praxis Tacticum: The Art, Science and Practice of Military Tactics*, which Chuck wrote a few years back.

> *Tactics is the art and science of organizing a military force with a specific goal in mind.*

EDUCATION, TRAINING AND DRILLING

These three words are frequently used interchangeably, but it's important to make a key distinction among them. The leader, or trainer, must appreciate the distinction among educating subordinates, training them, and drilling them. There may be significant overlaps here, but let's take a moment to consider why they are distinct actions, and why it is important to understand their distinctions.

Let's begin by defining our three terms in their most basic (military) forms:

- **Education** is the process of imparting knowledge, developing the powers of reasoning and judgment, and generally of preparing subordinates for the future.
- **Training** is the process of melding human and materiel resources into required military capabilities.
- **Drilling** is the process of repeating a particular action or methodology to practise military skills or procedures.

Because these processes are so closely related, they can easily be confused, and it's important for the leader, or trainer, to distinguish when and where to use one process over another. A good leader should inherently be an educator, whether it is in the classroom, in garrison, or in the field. This is more than a matter of teaching a particular lesson. It is also a question of being an example for subordinates to emulate. Training and drilling are intimately connected, but they are not equivalent. The simple reason for this is that although drilling is always training, training is not always drilling. This may not be obvious to the young leader, or inexperienced trainer so let's look at a couple of examples in a tactical context.

Let's consider an infantry platoon. The Platoon Commander wants to ensure that the platoon is proficient in the immediate actions required should the platoon be ambushed. There is a drill that forms the basis of these immediate actions, and so the Platoon Commander may task one of the Senior NCOs to prepare a short lesson on the ambush drill after which the platoon will go into the training area and practise the drill.

During the lesson, it may be education (for new members), or it may be a refresher. But clearly, once it deploys to the field the platoon is undergoing training since, to return to our definition, human and materiel resources are being melded into required capability. But what about the reverse? Can the Platoon Commander train the platoon on ambushes without drilling? Yes.

Let's assume the Platoon Commander wants to ensure that the platoon is proficient in the immediate actions required should the platoon be ambushed without using the ambush drill. This may begin with a lecture, but avoid talking about the drill. It may involve taking the platoon to the field and going through an ambush scenario without using the ambush drill. There are multiple options and reasons why the Platoon Commander would choose this type of training. Perhaps the platoon had spent the prior week drilling constantly using the ambush drill; perhaps the commander wants to force the platoon members out of their comfort zones to consider what to do in the absence of using the drill. Perhaps the Platoon Commander wants to demonstrate the importance of the drill by denying its use. Whatever the reason, it should be apparent that it is possible to train, without using a drill.

To sum up, in this case, the arrow of understanding only moves in one direction. ALL drilling is training; ALL training is education. But the reverse is NOT true. As a leader and as a trainer, your effectiveness depends on understating these fine distinctions and using them appropriately to benefit both your troops and you.

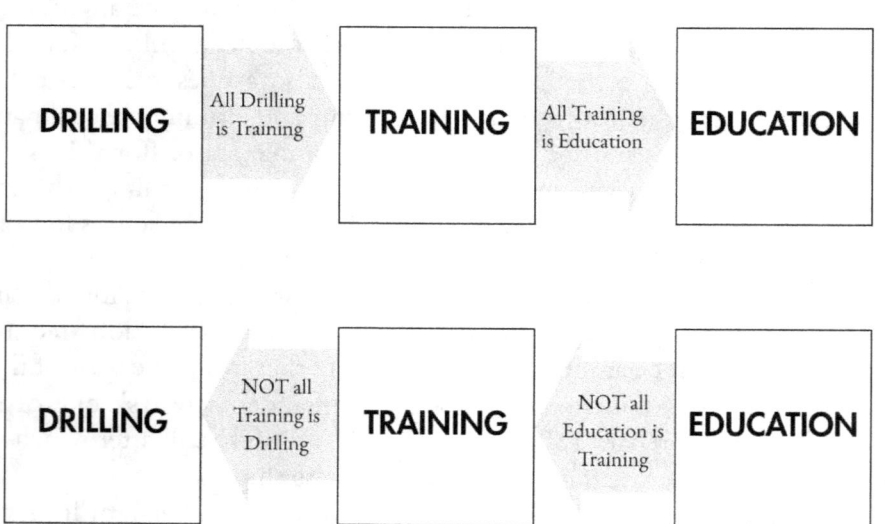

DRILLS: A MISUNDERSTOOD CONCEPT

During the Cold War, and even today when observing the ongoing Russo-Ukrainian War, the Soviet and post-Soviet armies have often been derided for their reliance on drills. A drill is simply a commonly understood and rehearsed reaction to a given situation. Because it has been rehearsed, it can be executed quickly and smoothly. The Soviet approach to drills was to use them almost exclusively as the form of maneuver for formations as large as battalion size. The "science" of military operations was held to be more important than the "art" that individual commanders might apply.

Soviet formations were assessed on the speed with which they could conduct drills rather than on finding innovative solutions to the problems that they faced. From a Western perspective, this made Soviet organizations seem overly rigid and predictable, liable to charge mindlessly into withering fire. Western armies tended to limit drills to very low levels, otherwise relying on commanders to use their initiative, accurately assess the situation in front of them, and determine the best course of action based on the circumstances. Every problem required a bespoke solution. We believed that this reliance on flexibility and individual initiative gave our forces a decisive advantage.

But the Soviets saw things differently. Their intention was not necessarily to rob commanders of initiative, it was to practice the easily predictable maneuvers of war until they could be conducted with ease – even under the extreme duress of heavy artillery fire, close air support, or within a chemical or radiation contaminated environment. The art that was left to the commanders was to determine which drill to use, where and when, so as to have the desired impact on the enemy. Drills didn't constrain them – they provided a common set of tools that a skilled commander could use in novel ways to achieve the aim. Thus, while in the West, drills were often seen as the antithesis of tactics, the Soviets saw them as the building blocks. Who was correct?

As with many things, the truth lies somewhere in between these extremes. An over-reliance on drills to the exclusion of all other possibilities would be too rigid, and lead to tactical mistakes. Equally, treating every tactical problem like a unique puzzle to be solved by the commander from scratch can lead to "analysis paralysis." Drills, when well-rehearsed and commonly understood, form a highly useful set of tools for a commander to use in a wide variety of situations. Where the drills don't fit, then they must either be modified on the fly, or discarded. But understanding them provides the basis for commanders at all levels to understand and thereby make good decisions. Drills are not tactics – but applying an appropriate drill for the situation is.

We now have another definition for our specific purposes. We can define tactics at the company level and below as:

> The art of taking the right action, at the right time, to achieve the desired aim.

But how do we know what "right" is? This shouldn't just be a matter of opinion, rank or hindsight bias. "Right" must be judged against a clear set of principles: those actions that create advantage over the enemy, preserve one's own force, and contribute directly to the mission. In other words, drills and bespoke solutions alike must be evaluated through a consistent lens — a way of thinking that allows commanders to recognise not only what can be done, but what should be done in that moment. Although leaders often struggle with this issue of what is "right" in training, where the consequences of decisions can be unclear, ultimately there is only one arbitator who decides if a tactical decision was the correct one: the enemy. In operations, it quickly become apparent which are the "right" decisions.

TACTICAL COMPONENTS

The various definitions above make it clear that some confusion exists regarding how to teach as nebulous a concept as tactics. The latest definition has the benefit of breaking the concept into easily understandable parts. Recall the comment about tactics being the purview of the senior officer present, or that what the boss thinks are "good tactics" is always right? Clearly this is not true once the enemy is given a vote, and this approach to understanding tactics doesn't create leaders who are capable of independent thought.

In order to teach tactics according to our last, and preferred, definition, we have identified four key components, each of which can be trained. They are: tactical acumen, tactical awareness, speed of decision-making, and tactical agility. None of these are more important than the others. Training must encompass them all in time, but not necessarily within the same training event. Below, as we describe each of the many different ways that you can teach tactics to your subordinates, we've mapped that method against its ability to teach those four components. Not all methods are equally useful for all components, and you will need to consider and weigh their relative merits when you plan your training. But teaching tactics should no longer be a nebulous concept — it is just a matter of training your subordinates in the component parts, as well as giving them an opportunity to put those components together into practice.

In order to not have to constantly explain the four components of tactics, here they are now, with their definitions.

Tactical Acumen
The ability to recognize the right decision to make, from all the available possibilities. This can be broken down into two parts:

- Knowing the possibilities, and
- Selecting the right one.

Knowing the possibilities is about having a thorough understanding of the available drills or actions, the capabilities of weapons and equipment (friendly and enemy), as well as understanding human nature. Acumen can be built through practice, and through historical study. Along with an understanding of the situation, this allows a commander to recognize the right decision.

Tactical Awareness
Also known as situational awareness, it is knowing where you (and your troops) are relative to everything happening around you. This obviously plays into tactical acumen as well, as the right decision is only "right" in relation to the situation. History is replete with examples of otherwise competent commanders whose plans fell apart because they were unable to maintain a clear picture of the situation they were facing. Without it, you are likely to be acting (or reacting) to situations that no longer exist.

Speed of Decision-Making
This is how quickly you can gather sufficient information and decide on a course of action. It's important to note that there are several ways to speed this process. The collection or transmission of information can be sped up, or the commander can become accustomed to making decisions with incomplete information. The use of drills can also speed up this process by providing a ready-made set of tools that can be applied.

Tactical Agility
This is the ability to rapidly change direction in reaction to an order or changed situation. Mental agility is built into speed of decision making, but tactical agility refers to the physical ability of your organization to respond. If you were advancing due north in a line abreast, how long would it take for you to reorient your force to advance to the east? Or due south? The ability to rapidly carry out orders or actions is key to an organization's tactical potential.

Tactical Agility vs Tactical Flexibility

There is a difference between tactical agility and tactical flexibility. This is more than just playing with words. The two terms are distinct and need to be understood. As noted, tactical agility is the physical ability of an organization to respond to a change in the situation. It is not just about the mental processes required to do so, but about the "nuts and bolts" of enacting a decision. In a practical sense, from the moment a commander has made a decision, there are still things that must occur before the decision is enacted. Subordinates have to receive the message, interpret it, consider their current and desired location, transmit instructions to their own subordinates, and physically change their disposition.

At the same time, leaders are often told to "stay flexible." Tactical flexibility is concerned with the leader's ability and willingness to be open to ideas that they either did not appreciate or did not come up with on their own. This relates more to tactical awareness, tactical acumen and the speed of decision making, and in fact could be considered to be the sum of these three things.

Quick Reference

DOES THIS METHOD TEACH:			
Tactical Acumen	Tactical Awareness	Speed of Decision-Making	Tactical Agility
✓ X	✓ X	✓ X	✓ X

To help you use these four components in designing your training, we have created a methodology checklist for quick and easy reference. By looking at the table, you can quickly decide whether the technique you would like to use to teach something is appropriate, and whether it will put emphasis on the right elements of tactics.

In each of the teaching methods, you'll be able to quickly see if what you are aiming for can be achieved using this method. This chart is not carved in stone, as you will see. It is important to note that on occasion you will see both a YES and a NO under one of the four components. We are not trying to be unclear. Depending on what aspect of the training technique you stress, the component will have a greater or lesser relevance.

OBSTACLES TO TEACHING TACTICS

One might assume that modern armies do little more than teach and practice tactics, but this could not be farther from the truth. Most armies do not explicitly teach tactics at all.

Why is this? Partially, it is because the concept of tactics is widely misunderstood. As was discussed above, tactics and drills often get confused with each other. There are no universally held definitions, and different military cultures have long-standing beliefs and customs, which sometimes may have an unintended negative effect upon training. As you will see below, all these factors combine to put constraints upon the teaching of low-level tactics.

Organizational Culture
One of the biggest obstacles to teaching tactics is rooted in the culture of many militaries. Generally speaking, there is too much emphasis on procedures over critical thinking. There's a strong focus on following standard operating procedures. Don't get us wrong — SOPs are essential. They provide structure and clarity, especially in chaotic situations. But when SOPs are treated as the "be-all, end-all," they can stifle creativity and adaptability. Leaders trained this way often become excellent at following orders but struggle when faced with a situation that doesn't fit neatly into the checklist.

Another cultural issue is the fear of failure. In many organizations, failure isn't seen as an opportunity to learn; it's something to be avoided at all costs. This creates a risk-averse mindset where both leaders and trainees are afraid to push boundaries or try new approaches. The result? Missed opportunities to learn through trial and error, which is often the best teacher.

Unrealistic Training Scenarios
Another obstacle is how training is conducted. Too often, training scenarios are idealized, pre-scripted, and predictable. These exercises might look great

on paper, but they don't reflect the messy, ambiguous, and unpredictable nature of real combat.

When everything goes according to plan in training, it's easy to walk away feeling confident. But what happens when you're faced with the chaos of actual warfare, where plans rarely survive first contact with the enemy? Overly sanitized training creates a false sense of security, leaving leaders unprepared for the complexities they'll face in reality.

Misconceptions About Tactics

There are also plenty of misunderstandings about what tactics really are and who should be learning them. Below, we have listed some of the most egregious.

- **The "Tactics Are Intuitive" Myth**. Some people believe that tactical decision-making is something you're either born with or not. It's seen as an innate skill that can't really be taught. This couldn't be further from the truth. Tactics are a teachable skill, and believing otherwise often leads to junior leaders being thrown into the deep end of the pool with nothing but their instincts to rely on.

- **Oversimplification**. Tactics are often boiled down to basic mechanical maneuvers or the execution of firepower. While these are certainly parts of the equation, they're only pieces of a much larger puzzle. Decision-making, situational awareness, and understanding the enemy are just as important—if not more so.

- **The "Tactics Are for Senior Leaders" Fallacy.** There's also a persistent belief that tactics are only relevant at the higher levels of command. This ignores the fact that junior leaders are the ones applying tactics on the ground. If they aren't trained to think tactically, how can we expect them to succeed?

Leadership Bias

Leaders can sometimes be the biggest obstacles to progress. They may stick to outdated methods or impose personal preferences on training, even if those methods no longer reflect the realities of modern warfare. They might also be especially sensitive to being seen to "fail," or even to have their subordinates fail. This can stifle good training.

Lack of Feedback Loops
Even when lessons are identified, they're not always fed back into the system. Without feedback loops, training programs fail to evolve, leaving armies stuck in the past and unprepared for emerging threats.

Institutional Inertia
Finally, there's the age-old problem of resistance to change. Militaries are, by their nature, conservative organizations. Long-standing traditions and practices often resist change, even when there's clear evidence that a new approach would be more effective.

All of these obstacles — cultural issues, unrealistic training scenarios, misconceptions about tactics, and resistance to change — combine to create a significant challenge for teaching tactics effectively. Addressing them requires a shift in mindset, a willingness to innovate, and a commitment to continuous learning.

THREE COMMON MISTAKES WHEN TEACHING TACTICS

There is a widespread myth outside of the military that only the best people are chosen for instructional duties. The reality is more mundane, as exposure to the training system quickly reveals. Instructors everywhere are chosen from whomever is available, based on the incorrect premise that anyone can teach military drills, procedures and tactics. When the best are available, then they are chosen, but frequently, those chosen to teach are run-of-the-mill individuals.

Occasionally, instructors are less than average and that is when errors, beyond the normal human ones, are introduced into what is taught. Once introduced, errors can be perpetuated from generation to generation of leaders and soldiers, becoming extremely difficult to root out. It's equally true that strong willed individuals in key positions can have an undue influence over training, erasing the results of hard-earned experience in favor of their own (limited) experiences or preferences. Over time, the error or misunderstanding becomes embedded in doctrine. That is the first step towards disaster.

In this section, we will look at the three most common errors that have become embedded in how many Western armies train. For the most part, these errors persist despite written doctrine that would suggest otherwise. Combined, these issues diminish the professionalism of the military and allow bad lessons to live on. Troops will fight the way they train. If you make these mistakes in training, then they will reappear in a real fight and there will not be an umpire nearby to reset the situation.

The three most common errors:
- Misuse of a Reserve
- Failing to Gain Benefit from After Action Reviews (AAR)
- Not Letting Subordinates/Plans Fail

Let's investigate each in turn.

Error 1: Misuse of a Reserve

All Western militaries have enshrined the need for a commander to maintain a reserve in their doctrine. In training it is common to not designate a reserve, or to create a "notional" organization to be assigned the task. The oft-heard excuse is that troops who are placed in reserve don't get to train — but that's an issue with how the exercise is constructed and the commander's tactical prowess more than anything else. A "canned" exercise where everything goes to plan often doesn't require the reserve to be employed. But with a few simple tweaks, the correct timing and employment of the reserve could easily be what brings about victory or defeat — as can often be the case in war.

Simply put, a commander must always maintain a reserve. More importantly, a commander needs to understand how to employ that reserve. Keeping a reserve is easier said than done, especially at the lowest levels of command. Platoon and section commanders often feel the need to commit all of their forces to their missions. But that doesn't remove the need to form a reserve whenever possible and to use it properly. Even a very small element that is kept out of an engagement so that it can be used to exploit an opportunity can have an outsized impact. A reserve is the commander's ace-in-the-hole. It's his tool to influence the fight once it has begun. Once all available troops are committed to a fight it becomes nearly impossible for the commander to influence the battle without using a reserve.

There are two recognized types of reserves:

- **Assigned**
- **Situational**

An assigned reserve is an element whose only task is to be the reserve. A situational reserve, however, is an element that is conducting other tasks, but will be used as the reserve only if required.

This widely held understanding contains the seed of disaster within it, because although these two types are commonly recognized, only the former is a real reserve because the latter is already conducting another mission. Relying on a situational reserve is fraught with risk. A true reserve has *no operational mission* other than to be a reserve. It sits ready to be called upon, but until then, it is *out of the fight*. Once a reserve is given a mission it is then — by definition — no longer a reserve. Consider the analogy of a cash reserve. If you have set aside a few hundred dollars for a rainy day and you use this money to buy new tires for your car or to take a holiday, there is no reserve left until a new one is reconstituted.

How a commander uses his or her reserve is the mark of tactical prowess. Poor or inexperienced commanders keep their reserves ready to save them from failure. That is *not* their purpose. The reserve is there for the commander to shape the tactical battle, to influence the course of events, and most importantly, to exploit success. There are many misconceptions and errors regarding the purposes and uses of reserves. Below we have compiled six of the most common (and most dangerous) along with brief explanations.

- **Failure to have a reserve.** This error is fundamental. Every commander *must* maintain a reserve. Any time you fight without one, you are rolling the dice with the lives of your subordinates. If you have no reserve, however small, then you have no way to influence the battle or exploit success.
- **Assigning other missions to a reserve.** This error is what was referred to above as the *situational* reserve, and as we have already said, it isn't really a reserve at all. Although it's common enough to be considered a type of reserve, it's not correct and should be avoided. It's a quick fix that too soon becomes the norm. When an inexperienced commander faces a problem that he feels requires more troops than he has, he will double-tap a subordinate with the extra task of being a reserve *on call*. This may play well on exercises, but the troops you've given two missions may simply not be available when you call for them. It is *not* a reserve; you are fooling yourself.
- **Using a reserve to avoid defeat.** This error is complicated. To the inexperienced tactician, this action may not seem so bad. After all, why have troops doing nothing when the situation is turning against you? Good question. Let's investigate. First, when a commander uses this tactic, he is not shaping the battle. He is reacting to his enemy. Second, in committing her reserve this way, the commander leaves herself unable to exploit any upcoming opportunity — the actual purpose of the reserve. So, what is a commander to do? There are multiple options depending on the situation, but generally speaking, the answer is to hold back the reserve while repositioning other

forces to shape the battle in preparation for using the reserve more positively. Easy to say; tough to do. All the more reason for you to study and learn.

- **Launching the reserve without reconstituting another.** This error is so common as to be invisible. In other words, most commanders don't see this as a mistake. This practice has become a common error because training exercises are short and there's an ingrained belief that once the reserve has been launched, the culminating point has been reached and soon it will be ENDEX. The reality can be punishingly different. An engagement can go on for many hours or even days. Failing to have a reconstituted reserve leaves the commander with no way to shape the battle or to exploit an opportunity that may arise. Apparent victory may devolve into defeat, for the want of a few more uncommitted troops. As Shakespeare famously said "For the want of a nail …"

- **Using a subordinate's reserve.** Often, a senior commander finds herself having launched her reserve and so reaches down and calls upon that of a subordinate. This is known colloquially as attaching a "string" to your subordinate's reserve, and it is *wrong*. If you have designs on troops that your subordinate controls, you need to task those troops as part of your orders process. In theory, if the commander reaches down and tasks a subordinate's reserve, he is now giving new orders to those subordinates. This practice is fraught with danger, and it runs the risk of destroying the necessary trust between leaders. Avoid it.

- **Attaching a string to a reserve.** As discussed above, if a "string" cannot be avoided, then a commander will put restrictions upon a subordinate's reserves because his own troops may be critical to the fight. He wants to ensure that more troops will be available if he (the superior) needs them. A commander may say, "you need my permission before launching your reserve." Fine, but in this situation what the commander is really saying is to the subordinate is "you are babysitting *my* reserve." That compels the subordinate to form TWO reserves, because if she can't freely employ it, then it is not *her* reserve. Again, this technique runs the risk of destroying the necessary trust between leaders. Avoid it.

Error 2: Failing to Gain Benefit from AARs

After Action Reviews (AARs) are deeply embedded in all NATO doctrine. They are not optional; they are requirements. Time must be built into training schedules for their conduct. There is no point in conducting training if you do not take the time to review how well the training was conducted. An AAR does not need to be formal. It does not need to be elaborate. It does not need to be long. But it *must* be done.

Unfortunately, AARs are often treated as optional or conducted in a cursory manner, leading to shallow analysis of outcomes. Skipping or rushing an AAR undermines its purpose, resulting in missed opportunities to identify and address critical issues. Additionally, discussions during AARs often focus too heavily on whether a mission succeeded or failed, rather than examining the decisions, processes, and factors that contributed to the outcome. This misplaced emphasis on results over processes limits the potential for meaningful learning.

Even when lessons are identified during an AAR, they are too often not applied. If you are able, take the lessons learned from the AAR and apply them immediately by conducting another iteration of the training, or at least a critical part of it. In this way, you confirm in the minds of the trainees that the lesson was valid, and it will be remembered. If the lessons cannot be implemented immediately, then there must be a commitment that they will be applied as soon as applicable. If not, they will never become lessons learned. A lesson that has been identified but not implemented remains of limited use.

By treating AARs with the seriousness they deserve, focusing on the processes that led to outcomes, and ensuring that identified lessons are implemented, leaders can ensure continuous improvement and readiness for future challenges.

Error 3: Not Letting Subordinates/Plans Fail.

It is an old truth that failure is a better teacher than victory. The sting of failure can be painful. But it can also be a powerful motivator to see what went wrong and to learn from it. Victory can breed complacency. Interestingly, even something that is less than a victory can be mistakenly thought of as a "win," so it is important for instructors at all levels to occasionally allow their students to live with their mistakes.

Naturally, allowing failure must be done with care. This important, but too rarely used, technique will only be effective if it is used positively, otherwise the trainees will not absorb the desired lessons. Doing this takes tact and excellent teaching skills. Trainees need to be shown that there is something positive to be salvaged from a plan gone wrong. Here is where a good AAR is invaluable. Use it to show your trainees that everyone — including you — can do better. Do not use the failure to assess blame; use it as a learning opportunity to become *better*.

Lastly, as an instructor you need not allow the failed action to proceed all the way to catastrophic defeat. Use your judgement and if you are at a juncture where it is obvious that the plan is failing, you may want to intervene, conduct an AAR, reset the training, and then continue. This reinforces several issues. You indicate to the trainees that failure was imminent. You demonstrate how changing the plan can snatch victory from the jaws of defeat, and you leave the trainees with a positive sense of accomplishment, the final win erasing the earlier defeat.

ORGANIZING TRAINING

The importance of understanding the purpose of a training event cannot be overstated. It would astound most leaders to discover how often troops deploy for training with only the vaguest notion of why they were doing it.

The overall purpose of every training event is made up of the same three components:

- Aim,
- Scope, and
- Training Objectives.

In other words, what is the ultimate aim of the training, what are the parameters that define how big or small the event will be, and what specific things is the training supposed to achieve? Whether it is a one-hour tactical problem or a ten-day field training exercise, this triad must be understood by the person in charge of the training and communicated to their subordinates.

Furthermore, the leader needs to appreciate another key component of the training process: who the primary training audience (PTA) is. At the squad, section, platoon or company level, the PTA will almost certainly be your soldiers. But there are possibilities even at this low level to make a fundamental error that could derail your training. Let's have a brief look.

Primary Training Audience (PTA)
By definition, the PTA is the organization to be trained. This organization must be in a single echelon of command. They are the people who will *directly* benefit from the training. What this means is that the PTA cannot contain members in different echelons of command. For instance, if you are a company commander, and you want to train your platoon commanders, they are the PTA, and the training must be focused on *them*. In the process of training the PTA, other soldiers (such as those within their platoons) who are partaking in the training will gain benefit, but it still *must* be focused on the PTA. In this example, the other soldiers, who are not platoon commanders, are thereby considered a Secondary Training Audience (STA).

Secondary Training Audience (STA)
Fundamentally, the STA is anyone *not* in the PTA. Let's refine that a bit. The STA comprises people undergoing training along with the PTA, but for whom the training was not *specifically* designed. They derive benefit by being part of the training, but they are not the focus. Their benefits are incidental because the training was not designed for them. This doesn't mean that they will receive poor training, just that they are not the focus. The inexperienced trainer may be tempted to amend the plan to give more benefits to the STA — and this is a mistake. Never change the training unless you are doing it to enhance the training of the PTA.

Differentiating PTA and STA
It's not always clear who is the PTA and who is the STA, and it can get a bit complicated, so let's break down this concept with an example. Imagine that you are a company commander, and you have been directed to take your company into the field to conduct one day of defensive operations training in advance of a battalion level exercise. You know that the battalion exercise will see you digging in a defensive position and defending it over the course of several days.

What training can you conduct with your company in one day? The obvious (and wrong) answer is to take your whole company into the field and have them dig in.

Let's start at the beginning. What is the aim of your training? What specifically do you want to practice in advance of the battalion exercise? And who is you PTA?

The aim of this one-day exercise should be to focus on some critical element of an infantry defense for the battalion-level exercise. This is not about practising exactly what you will do on the battalion exercise so that you and your company look good. It is about training aspects of the task that are important, and difficult or impossible to train when you are no longer the person running the exercise.

For the sake of this example, you decide you want to focus on a problem that you identified in an earlier exercise — a lack of coordination amongst your platoon commanders. This becomes your training aim.

The outer limits of your scope has essentially been given to you already — up to a day in the field, with up to your entire company. But if the focus is on your subordinate commanders (as the PTA), what do you do with the rest of the company? Do they even have to deploy to the field at all?

That depends on how you construct the training.

You might determine that the key training objectives you want to practise are the siting of trenches with interlocking fields of fire and the sequence of occupying a defensive position. These are the training objectives for your PTA. You could lead your subordinate commanders to multiple pieces of ground, having them site their trenches on each one and then critique each other's selections. By conducting an AAR after each position, you can quickly incorporate the lessons learned into the next iteration. The activity on each defensive position would take no more than an hour to conduct, allowing you to easily do multiple iterations of this training in a single day.

There is no need to involve the rest of the company in this activity, unless there was a reason to have them observe. (In which case they become the STA, by the mere fact of observing.) You certainly would not want them to start digging in each position, only to be moved to a new one just as they got started. But you could have your section or squad leaders take the remainder of the company and conduct training on how to properly construct a trench, the sequence of occupation of a defensive position, or any other areas that you feel are critical. That separates them from you, so there is no STA. But let's return to the earlier possibility.

THREE COMMON MISTAKES WHEN TEACHING TACTICS | 23

You decide that halfway through the day, you would like the remainder of the company, controlled by the platoon sergeants, to come and participate with their platoon commanders. They could nominally "occupy" the various platoon positions that you went through with the platoon commanders, so that they could learn how the positions evolved. The platoon commanders thereby have their lessons re-enforced by putting their troops through the processes, and the soldiers gain insights into the siting of platoon positions. This process has made them an STA.

There are many other combinations of training made possible by changing the objectives and PTA/STA, all of which would achieve different outcomes without markedly different resources being involved. Your creativity as a leader, as well as your clear-eyed view of what you are trying to achieve, will make all the difference. Here is the bottom line: the PTA is who the training is aimed at. Everyone else slips into the STA.

PART TWO
SIMPLE TRAINING

METHOD 1 - SAND TABLES

DOES THIS METHOD TEACH:			
Tactical Acumen	Tactical Awareness	Speed of Decision-Making	Tactical Agility
✓	✓	✗	✓

Most military units allot certain times during the week when leaders are expected to discuss matters of importance. These could include unit policies, finance, leadership and so on. If you are responsible for one of these Leader Sessions or Commander's Hour, tactics should be one of your recurring topics. Let's look at how you might use a sand table to expand your troops' tactical knowledge.

This form of training uses a piece of simulated terrain, the "sand table," and models or markers to represent units, obstacles, and features. Although much less common than in the past, a sand table remains an excellent methodology. In its original form, it is simply a sandbox mounted on legs that allows an instructor to quickly shape a piece of terrain with hills, low ground, etc. Instead of a sand table, the same effect can be achieved by using a whiteboard, a large sheet of paper with terrain drawn on it, chalk on the size of a vehicle, or even a PowerPoint slide or Miro board. The key element of all of these is that the instructor can quickly and easily create a simple representation of a piece of terrain.

Although in a traditional sand table, models are used to represent vehicles or units, this is not the only solution. Plastic soldiers can be used, as can blocks of wood, tented cards, or sticky notes. The important thing is that each element is clearly visible, well-defined as to what it represents, and easily movable around the terrain.

The sand table and models are used to represent a simple scenario, typically a moment in time when a drill must be executed. This could be something new that you wish to instruct on prior to trying it in real life, or it could be a drill that you are reviewing in case it's needed. It can also be used to teach when to select a drill, or how to modify it "on the fly."

A sand table exercise differs from a "rehearsal of concept" or ROC drill in several ways. Firstly, it is meant to examine a simple, discrete concept rather than all the elements of an operation. Secondly, subordinate commanders do not necessarily brief their own actions during a sand table exercise, as they do during a rehearsal — the method is much more fluid than that, and anyone can be asked to brief any aspect at will. [2]

This method can also be used with experienced troops to discuss a potential modification to an existing drill, and how you want them to react. For example, if a drill calls for a platoon to shake out in line abreast, but the terrain is too restrictive to allow this, what do you want them to do? Should they narrow the spacing between elements of the platoon to fit into the available space? Or should elements fall back to form a box or other formation, to maintain the normal spacing? By walking your soldiers through a modified drill using the sand table method, you can ensure smooth execution of the drill in real life, even in imperfect conditions.

This method can also be used virtually. A virtual whiteboard or similar online tool can be used as the "sand table," and markers on it used in place of the models. Participants can move the markers, or the instructor can do so for them if necessary. Even sharing a screen while using shapes on a PowerPoint slide can achieve a workable effect and allow the sand table method to be used. The only limit is your imagination.

How Does This Method Teach the Different Components of Tactics?

Tactical Acumen

The sand table method can be used to present situations that force modifications to an existing drill or procedure, or even teach a tactic. Typically, there are multiple solutions to every problem, some better than others. An example of this noted above is when terrain narrows and forces an organization in line abreast to modify its formation. What is the right answer? It depends

---

[2] See the Special Note at the end of this chapter.

on many factors. Discussing these factors and the reasons why one decision would be made over another is well suited to a sand table session.

Tactical Awareness
A large part of tactical awareness is developing the ability to "see" the situation clearly. The sand table method allows the participants to visualize complex situations in a simplified way. Particularly for junior leaders, they may be isolated from their parent organization during the execution of tasks, and may not have a clear idea of what every other element is actually doing. This method allows them to create mental pictures that will carry over to live execution.

Speed of Decision-Making
Unless forced by the instructor, this method is not ideal for practising rapid decision making. This is because it is better suited for slowly and deliberately walking through a tactic, technique, procedure (TTP), drill or situation, ensuring that it is clearly understood at each step.

Tactical Agility
This method is ideal for confirming or reinforcing the details of any event, showing participants each step and allowing them to see how all the parts fit and work together. It is less suited for training the critical decision of which action to implement at any given time, as it lends itself best to very simple scenarios. Other methods that model more complex scenarios are better used to practice this skill.

Critical Elements of the Method

Keep a Sand Table Kit Handy
The beauty of this method is that it can very quickly be used to conduct training that is tailored to the situation that you are facing or are about to face. A leader can keep a Ziplock bag with the basic elements to use this tool in their field kit, ready for use. A bag of plastic soldiers (or any other markers) and a few pieces of mine tape are enough to be able to create virtually any scenario. You can tailor your kit to your organization, so that you have markers for each of your vehicles, key personnel, etc. The only limit is your imagination, though simplicity remains a virtue.

Question Rather Than Lecture

Using this method to simply demonstrate an action might be necessary with troops who are untrained or unfamiliar with basic tactics, but a trained audience should not be lectured. Instead, have subordinates describe each step of the drill and move the related markers. This confirms their understanding of the drill and can be used to reinforce elements of it with the subordinates who will play pivotal roles in the drill by having them speak to their parts, or present novel problems that force exiting procedures to be modified and draw potential solutions from the audience.

Have the Reference Nearby

The value of drills comes in part from the fact that the interoperability of a force is increased when there is a common understanding of how drills are executed. Often, units develop variations to common drills, sometimes based in a drive for efficiency and sometimes based on other, less relevant, factors. When teaching or rehearsing a drill, having the relevant reference to ensure that all aspects of it are taught and practised correctly can greatly increase the value of training. An instructor teaching a half-remembered (and possibly bastardized) drill only increases confusion.

Challenges When Using this Method

Keep it Simple

The sand table method allows you to abstract many unnecessary details so that you can focus on the essential elements of the drill or tactic. Don't fall into the trap of adding this detail back in to satisfy the curiosity of the participants. Rather than answer questions like "what's the scale?", make decisions that give the necessary information but no more: "the spacing between vehicles here is 100m," or "the enemy is within effective range."

Avoid Rabbit Holes

All forms of training suffer when the aim is lost, and this is equally true of sand table exercises. Because the setup of this form of exercise is so simple, it can be easy to add new elements to the problem that distract from the intended purpose. Maintain focus on what the intention of the training is, and "park" interesting but irrelevant detours for later training.

PRACTICAL EXAMPLE 1

You are a platoon commander. Next month there will be a force-on-force exercise, and you know from experience that there will be at least one ambush during the exercise. You decide to use your weekly Commander's Hour to do a sand table exercise with your whole platoon.

After you have gathered everyone around your sand table, you call upon the platoon sergeant to lay out a platoon-level ambush using the sand table (you have already warned him off, so he is not surprised). While he lays out the plastic figures on the table, you tell the platoon the aim, scope and training objectives of the sand table exercise:

- **Aim**: review the ambush.
- **Scope**: lay out an ambush at the platoon level in isolation.
- **Training objectives**: refresh everyone's memory; demonstrate actions and reactions during the conduct of an ambush; ensure that squad leaders are fully able to conduct an ambush in the absence of the platoon commander and platoon sergeant; and confirm that every soldier is comfortable with their role in an ambush.

While the sergeant finishes building the model, you ask questions of the squad leaders. What is the purpose of an ambush? Give me one critical component of an ambush. Why do we always leave one soldier out of an ambush? These questions are only intended to get the soldiers in the right frame of mind until the sergeant has prepared the model.

Once the sergeant is set, you ask him to describe the various components of the ambush and their functions, moving them as needed to illustrate his explanation. Here might be a good place to stop for questions. Continue the activity until the training objectives have been met.

This is a rudimentary example of how you might use a sand table, using only an hour of time. In order to be successful, ensure that you prepare yourself and your second-in-command, know what you want to achieve, and do not waste time.

PRACTICAL EXAMPLE 2

You are a section commander, and your section will be relieving another section on a week-long perimeter security detail. Your 10-soldier section is on rest in preparation for relieving the on-duty section in 12 hours, so you decide to take an hour and perform a sand table exercise. Your section has rotated in and out of this duty twice this week so they know what to do, but you would like to discuss what would happen if the small compound came under attack, while you and your soldiers are on shift. You gather the section around your armored vehicle and using chalk, draw a schematic of the compound. You ask your second-in-command to draw where soldiers will be and while she is doing that, you tell everyone what you're doing.

- **Aim**: discuss defense of the compound.
- **Scope**: fighting off an attack as a section without any outside support.
- **Training objectives**: re-enforcing immediate actions; communications procedures; first aid procedures; calling for assistance.

As with Practical Example 1, you need to be prepared. You need to know what it is you want to achieve, and you need to communicate your intentions to your subordinates. It doesn't have to be a formal event. As a section commander, you are only one step removed from your subordinates, and it can be relaxed. But you must maintain control and it must be a positive experience for all involved.

After that, keep an eye on the clock and carry on.

SPECIAL NOTE ON BRIEFINGS

One of the many overlooked skill sets for young leaders is learning how to give a briefing. The key is in the title: Be brief!

Giving good briefings does not come naturally to most people; it must be taught. Traditionally, this skill is taught at junior staff courses for officers and leadership academies for NCOs. As a leader, you can (and should) begin teaching this skill to your subordinates as part of their tactical education. Every junior leader, whether NCO or officer must be confident and capable of briefing troops or senior leaders at the drop of a hat. Each army has its own preferences on how briefings are to be given, so we won't go into the mechanics of a briefing but teaching your subordinates a few basics will ensure that you ease their nervousness when you ask them to brief their tactical decisions and considerations and also assist them in imparting vital tactical knowledge to their various audiences.

The single most important aspect of any briefing is that the briefer needs to know how much time he or she has. We've all sat and listened to "briefs" that went on for twenty or thirty minutes and began with those famous words, "I'll be brief." Train your subordinates by routinely asking for briefings on mundane subjects that they know something about. "Jones, brief me on your LAV. You have 4 minutes." Jones will stutter and be nervous but no matter what he says, at the 4 minute mark EXACTLY, stop him. Thanks Jones, your time is up. Do it again to the next solder but give them 3 minutes and 15 seconds then do the same. DO NOT be nasty. Simply point out that your timing is a HARD STOP. Over time, you can change the subject as you work your way through the platoon. ALWAYS change the timings and keep them seemingly arbitrary. Your soldiers will rise to the challenge! They will soon be able to rise confidently and give a 2 minute 17 second briefing on why tracks are better than wheels or on why a killing burst from a machine gun should not be longer than 10 rounds (or whatever).

NOTES

NOTES

NOTES

METHOD 2 - FIELD PROBLEMS

DOES THIS METHOD TEACH:			
Tactical Acumen	Tactical Awareness	Speed of Decision-Making	Tactical Agility
✓	✗	✓	✗

You are deployed in a training area and have been told by your commander that your unit has one hour of unexpected free time. You don't want to tell your soldiers to stand down. Instead, you want to use the time to teach tactics. How to proceed on such short notice and still maintain a professional atmosphere?

One of the simplest ways to teach tactics is with what we'll call Field Problems, which require no resources other than time and imagination. In this form of exercise, you as the leader use whatever ground is immediately available and which can be viewed by the training audience. The focus here is less on finding the one "correct" solution than it is on generating possible solutions and examining them to tease out learning points.

Simplicity is key to this method of teaching tactics, and it's based on three components. Modifying any or all these components can create very different training outcomes, and so you need to be careful when devising the problem. Let's look at them individually.

The Situation

You should present a simple tactical problem without any additional information that might *seem* necessary, but will often only obscure the true nature of the problem. Don't give orders. Typically, you should offer a simple description of the enemy's size, disposition, and aim. However, depending on what you want to achieve, you could leave out some of this information, as is so often the case in war. Don't take more than 5 minutes to present the problem. Keeping the details to a minimum will help focus the training. You might also want to emphasize that there is no perfect solution, and that the aim is to generate discussion on the way to finding a solution.

The Question

How you phrase the question will determine the course of training. It's at the heart of what you want your students to absorb and the tactical problem they need to solve. Your question is best expressed in plain language, in the most unambiguous way possible. Typically, you would begin with something like: "How would you …?"

The Outcome

This is the third leg of the stool, the "product" the students need to produce. This must be simple. Ideally, it'll be verbal, though it could also be in the form of a sketch. You could ask students (if they are leaders) to deliver a verbal Fragmentary Order, just as they would give over the radio. It could also be a short briefing, no more than two or three minutes long, which uses the ground itself for reference. Whatever the form, you must be explicit about what you expect when giving instructions to your students.

The problem should be aimed at the level of the training audience, but since you are keeping it simple, there's no reason why junior soldiers can't be given problems that are "above their pay grade." Do this occasionally to challenge them and promote their use of initiative. You can also use this "stretching" technique to give them a window into what their superior commanders are doing and thinking.

After you have stated the problem, they'll need time to consider it. As a rule of thumb, set aside up to half of your remaining time for them to consider a solution. Once the time is up, you can either select someone to brief his or her solution or open the floor to general discussion. Whichever you choose, begin by reminding everyone that there is no single solution, that there are many ways to skin this tactical cat and have them give their solutions in reverse order of rank, the junior person going first. This will help to prevent deference to senior opinion.

Do NOT offer a solution of your own.

Allow free discussion but keep a tight rein on it. Do not allow anyone to drag the discussion down a rabbit hole or introduce a topic that takes you away from your desired aim of discussing this specific tactical solution. If that happens, interrupt and put the discussion back on track. Allow this freewheeling for as long as you can. Remember to praise volunteers and those offering comments and positive criticism. Remind everyone that you are not looking for any single solution; you want to hear their thoughts and opinions. If the discussion lags or everyone seems to have settled on a single solution, you can ask "Does anyone have a different solution?" to keep the discussion alive.

Once you feel that you have sufficiently milked the discussion, it will be time to wrap up. Summarize the key points that have come up in the discussion. Praise the group for what they have accomplished thus far. Tell them you know that you are pushing them, but it is all part of the education process. Time is everyone's enemy, so thinking quickly is a necessary tactical skill. It's important that you, as the leader, do not offer a solution. Doing this will quickly stifle creative thinking amongst the group, and lead them to the erroneous belief that there is only one right solution to the problem. It is hard to overstate the stifling effect that a leader can have when they express their thoughts on a matter under consideration.

Review the various solutions and comments. Tell everyone it was an excellent discussion. Tell everyone that you learned something and thank them for their engagement. Note: Don't discount the possibility that one of your subordinates will have learned an important lesson that you had not intended to teach. When that happens, do not be discouraging. Be positive. Praise those individuals for looking at the problem more creatively and for drawing more meaning from such a short lesson.

These instructions may seem overly focused on praise and positivity. This style of freewheeling discussion, where people offer their thoughts and solutions in a "free market" of ideas can be very intimidating, and runs counter to military culture. You will need to use encouragement to create a culture where everyone in your organization is comfortable expressing themselves in this manner, as well as comfortable dealing with each other's thoughts in a positive and constructive manner. Once this culture has been built, it must be maintained. All the while, your subordinates' natural instinct will be to determine whether they are "right," and if they came to the same conclusion as you. In the absence of this sort of validation, positivity is key.

How Does This Method Teach the Different Components of Tactics?

Tactical Acumen

These kinds of activities are simple and inexpensive means to draw upon the full breadth of the tactical knowledge of your students. Because the problem that you pose is only limited by your imagination, any situation can be conjured up and any aspect of the conduct of war can be considered. By drawing out solutions from a wide variety of people, there is also a greater likelihood of finding novel or creative approaches to the problem.

Tactical Awareness

This aspect of tactics is not fully trained using this method, as a key element of it is about recognizing changes to the situation as it unfolds. This method does develop the students' "coup d'oeil," however, especially if they are given only a short time to develop a solution. One possible way to focus on this aspect of training is to interrupt the students while they are developing their solutions to inform them that "the situation has changed." Then give them new information that upends their previous thinking (or appears as if it should). But be careful. If not used judiciously, this technique can derail the exercise by sending students into a tailspin.

Speed of Decision-Making

By setting a short time limit for the creation of solutions to the problem you have posed, speed of decision-making can be trained very effectively. It can be interesting to see the variety of solutions that are generated in a short period of time versus a longer one. Often, a longer period of time to consider the problem will not generate better solutions than a shorter period of time, especially when the students have experience with similar problems in real life.

Tactical Agility

While this method can be used to teach mental agility (as part of speed of decision-making), it doesn't teach tactical agility because there are no actual troops being exercised. Changes to disposition or direction of travel are all within the students' minds, and are essentially instant.

Critical Elements of the Method

Don't Fixate on a "Right" Answer

As an experienced leader, you likely have a solution in mind to the problem you have posed before the activity begins. This may be the best possible solution, or you may be surprised by a better one. It's important that you not attempt to steer the discussion around to accepting your solution. This defeats the aim of using this technique. Instead, use the student responses to explore the problem fully. Equally, if you are an inexperienced leader, you don't need to know the answer to the problem you're posing for this exercise to be effective. You can draw answers from more experienced subordinates, and through questioning and discussion, determine which are the best solutions.

Determine the Pros and Cons

There is never a single *correct* solution, so the discussion should focus on determining the pros and cons of each solution offered. This investigation can lead to interesting an informative discussion if you let it, but you need to manage it well. For example, one student's solution may have a small reserve, while another may have a larger one. By comparing the impact of these options and exploring how these may cause the plans to unfold, a deep discussion about the utility of reserve forces can be had. A canny leader will identify these opportunities in the discussion and exploit them when they occur, as students may bring out aspects of the problem that are unexpected.

Reward Creativity

Successful tactics require creativity, but the military approach to training and briefing can easily stifle it. As a leader, you must work to create an environment where students feel free to express creative solutions to the problems they face, even if their solution might appear "wrong." Subordinates should be praised for creativity, even if through discussion it's determined that the cons outweigh the pros of their solution. Time permitting, you can use an unusual solution to get the students to analyze what the flaws in the solution may be. Often, more is learned from analyzing a failure than from a success. This is not the same as rewarding bad answers – finding creative solutions, and then weighing them to determine their worth, is a key part of effective military planning.

Use Probing Questions

If you merely have students present their solutions without asking probing questions and generating discussion, the utility of this exercise will be limited. Most of the learning in this activity isn't achieved by simply generating answers. The learning takes place in defending and discussing them. This is the heart of the exercise and should therefore use the greatest share of the available time.

Challenges When Using this Method

Overcome Subordinates' Desire for More Information

Some students, typically at more senior ranks, will insist that they require more information to provide a solution. Too often at advanced schools and staff colleges, students are purposely given overwhelming amounts of information as well as resources as part of a tactical problem. This technique is aimed at teaching senior leaders to quickly sift through mountains of detail to come to the heart of the matter. It's not what we need to do at more junior levels. Although key questions can be answered as part of the initial briefing, you must resist the temptation to be distracted from the essential elements of the problem. "What weapons do the enemy have?" "How much fuel do we have?" "How heavy are the enemy vehicles?" "Is the weather good or bad?" None of these questions really address the heart of the problem being discussed, and you need to encourage your students to "pull up." In other words, focus on the essential and ignore unnecessary detail – even if that detail is interesting.

Manage the Discussion

You need to strike a balance between free discussion and stifling creativity. This comes with practice but generally, it's better to encourage free speech than to stifle it. Don't allow anyone to sidetrack the discussion (Is the enemy in BMP-1 or BMP-1A?) or introduce a topic that takes you away from your desired aim of discussing a specific tactical solution. If that happens, you need to interrupt and put the discussion back on track. Particularly if you are an inexperienced leader with more experienced subordinates, you may find it challenging to manage the discussion. Allow your subordinates to explore issues that are of interest to them, and to question each other, but don't spend all the available time on issues that are not critical. Using your more experienced NCOs to assist you is one way to do it. You may be surprised at what comes out of the discussion, but don't allow your subordinates to take over.

There Will Not be a Single Solution

Sometimes, at the conclusion of the exercise, subordinates will ask you for "the answer." We're accustomed to their being a "school solution" to tactical problems, or to expecting the senior person present to declare what the "right" answer is. It's for this reason that we often hear the old saw "tactics are the purview of the senior officer present." You should avoid this at all costs. You can highlight that some answers had more pros than cons, and vice versa, but that all the solutions have their utility. Battles are often won by the quick and violent application of "imperfect" solutions. As armor commanders are often taught: an imperfect solution now is better than a perfect solution offered too late. Be flexible.

PRACTICAL EXAMPLE 1

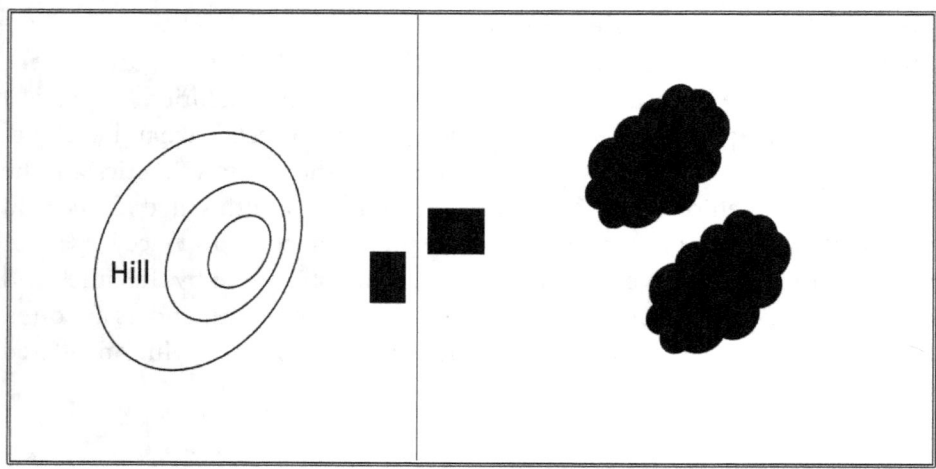

The sketch map above represents a simple piece of terrain where you are currently located. North is to the top of the sketch. Two buildings sit astride a road running North/South through the center of the sketch. To the West of the buildings is a hill, and to the East are two copses. This sketch represents the ground where the Field Problem is being executed.

You determine the following structure for your training:

- **Aim**: discuss an attack at the platoon level.
- **Scope**: a platoon, operating independently.
- **Training objectives**: use of terrain, attack drills.

The instructions you give your participants can be as simple as this:

> **An enemy section is occupying the two buildings. You are commanding an infantry platoon in the North copse. The enemy mission is to block all traffic on the road. How would you attack them with your platoon? You have thirty minutes, and then I will call on you to provide radio orders to the platoon.**

PRACTICAL EXAMPLE 2

You are standing on the same piece of terrain as in the first practical example. By changing the structure of your activity, however, you can train entirely different elements of tactics. You decide instead that:

- **Aim**: discuss a defense at the platoon level.
- **Scope**: a platoon, operating independently.
- **Training objectives**: flexibility, use of terrain, defensive drills.

The instructions you give your participants can be as simple as this:

> **You are commanding an infantry platoon occupying the two buildings and you're your mission is to block all traffic on the road. An enemy of unknown size reported to be moving rapidly along this route from North to South. You have five minutes, and then I will call on you to give a two-minute briefing on the plan for the block.**

Then, at the five-minute point, you might announce:

> **The enemy is NOT coming from the North; it is coming over top of the hill to the West. You have another five minutes, then prepare to brief me on how you will execute the block.**

Time out for a story to illustrate a point: Years ago, I was an instructor at the Army Staff College. We had a fighter pilot in the syndicate and when faced with the tactical problem of how to counterattack a Soviet Independent Tank Battalion that was proceeding through a town and across a water obstacle to our main defense, he opted to leave a friendly tank battalion behind, hidden in the underground parking garage of a mall. The solution generated riotous laughter and was dubbed "The Scotty Plan". Everyone, including me, agreed that it was likely suicidal.

Upon reflection, I decided to "test" his solution, so I grabbed the syndicate, and we drove the 45 mins to the town to walk the ground and allow the young captain to defend his plan. Although far from an ideal solution, with suggestions from the syndicate combat arms students and some lively discussion, we all eventually agreed that the plan was audacious and might create enough casualties and havoc for the friendly force to do its job and successfully return.

The bottom line: We had a full day of tactical discussion that was creative, interesting and morale building.

NOTES

NOTES

NOTES

METHOD 3 - BATTLEFIELD STUDIES

DOES THIS METHOD TEACH:			
Tactical Acumen	Tactical Awareness	Speed of Decision-Making	Tactical Agility
✓	✗	✗	✗

A battlefield study is an excellent way to teach tactics. And as with many of the methods in this book, the possibilities are limited only by your imagination. As a junior commander, do not feel constrained to use only historical events at your tactical level. There is nothing wrong with taking an infantry company to study a battle that featured a clash of major formations. No matter how large or small the forces were in the case you selected, there will always be something valuable to teach your subordinates, whether they are recruits, experienced NCOs or junior officers.

If you live close enough to an historical battlefield, you can take your troops to the site to walk the ground and investigate what happened. If that isn't possible, you can study the chosen battle using maps, terrain models and videos. (YouTube has many videos of reconstructed battles, but use them with discretion – not all are accurate.) Google Maps can allow very detailed investigation of the ground, especially if "Street View" is available. If you are fortunate to have a military simulation center nearby, you might also consider building the historic battle in simulation. It will not necessarily have to be "fought" by your trainees, as it could also be allowed to run with both sides under computer control, and the results viewed afterwards as a "playback."

In the past, studying history was the preferred method for soldiers to learn their trade. History's great military leaders, from the ancients right up to the 21st century, all credit their success to their study of past battles. This is not because history repeats itself — it doesn't. But it does rhyme. Although this aphorism is often attributed to Mark Twain, it is actually from a 1965 essay by psychoanalyst Theodor Reik. This "rhyming" gives us the opportunity to study a previous battle to learn tactics that may help us in a future one.

The key question is: How do we best go about doing so?

The first step, as always, is to decide what exactly you wish to teach. Remember: aim; scope; and training objectives. With all of history at your fingertips, the choices are endless. The key is to keep the number of lessons small. Do not be overly ambitious. Keep it simple so that your chosen lesson does not get lost in the detail of the battle. For instance, Gettysburg can be used to teach any number of lessons. DO NOT choose too many.

The time period in which the battle occurred is less important than you might first think. While an ancient battle might seem to have little to offer the modern student, it is important to see beyond the superficial details of the weapons used, and instead focus on the aspects of battle that are enduring — such as the principles of war, the impact of leadership, and the human factor. By focusing on these aspects of a conflict, thoroughly modern lessons can be drawn from any time period.

In order to ensure a clear understanding of a conflict, and allow for relevant lessons to be drawn from it, you may need to spend some time examine the details of period weapons and tactics. Be aware that this study can easily become very time consuming, and obscure the intended training. Ensure that whatever historical "deep dive" you take remains in service of the wider aims of the exercise, and when in doubt, recognize that less if often more. In other words, keep it simple and stay focused on your training aim.

Consider your teaching aids. They will depend on whether you are physically going to walk the ground or whether you will study the battle at a distance. Using simplified sketches and drawings is always a good idea, but have at least one or two tactical maps to keep the reality of what actually happened available for discussions and analysis.

Both you and your subordinates will need to prepare in advance. Unless you a highly skilled instructor (in which case you are probably reading the wrong book) you will have to create a lesson plan laying out *what*, *when*, *where*, *how*, and *why* of what you will be saying. Similarly, you will need to give your subordinates some reading material to familiarize them with the battle that you will be investigating. This will help them contextualize what you say from the onset, rather than learning for the first time what went on. At a minimum, prepare a handout with a broad-brush overview of what happened, a timeline and an indication of what tactical lessons they will be focusing on.

A very effective method that can be used with more senior students is to assign them to take on the role of different historical commanders who were present during the battle. It is best if you provide them with research materials that you have reviewed to ensure that they contain the necessary information, rather than asking them to conduct the research on their own. Once assigned, you can then call upon these students to present "their" commander's point of view during various points in the battle, explaining their aims and concerns. This helps the overall battle to come alive, and places your students in the thick of it.

How Does This Method Teach the Different Components of Tactics?

Tactical Acumen
Depending on the time period of the battle selected, the tactics employed may be familiar to a modern soldier or very different indeed. This need not have an impact on your ability to use the activity to teach tactical acumen. No matter the details of the tactics, many of the decisions that commanders face will feel familiar. Whether to advance or retreat, attack or defend, commit a reserve or hold it for a more opportune time — these are all timeless dilemmas. By making these elements your focus, effective learning points can be pulled from the most unfamiliar of situations.

Tactical Awareness
Because of the nature of a battlefield study, the whole of the situation is generally known to participants before the activity begins. As a result, it is difficult to use this method to teach tactical awareness.

Speed of Decision-Making
Typically, participants in a battlefield study are not expected to make decisions — instead they are examining the decisions made by others. This method is not well suited to teaching speed of decision-making.

Tactical Agility

Since this method is largely retrospective, and there are no physical forces to maneuver, there is no opportunity to practice the tactical agility of the participants.

Critical Elements of the Method

Use History to Highlight Foundational Ideas

It's important to remember that during this form of training, history is the method by which you are teaching tactics, rather than being the end itself. The primary aim of this exercise is not to teach history, or even historical tactics. Instead, use history as the tool to highlight timeless ideas such as the principles of war, or other concepts that are relevant to a modern audience. What they learn about history is a bonus.

Detailed Preparation is Key

A battlefield study requires research ahead of time to ensure that all the necessary information is in hand, and the participants have had time to review it. There will always be students who are avid historians who either conduct additional reading on their own or come armed with a knowledge of the battle you are studying. These "buffs" can quickly dominate the activity unless all the participants have a sufficient grounding in the scenario you are examining. Ensure that you provide them with enough information to fully participate and draw out the kinds of observations necessary to make the activity a success.

Challenges When Using this Method

Don't Get Lost in the History

History is fascinating, and when it is combined with the experience of actually walking the ground where a battle occurred, it can be very distracting. It's important to remember your training objectives during the conduct of a battlefield study, and guard against getting lost in interesting but otherwise irrelevant historical details. Ensure that the history serves the purpose that you have set, and recognize when the historical detail you have is sufficient for that purpose.

Misunderstanding Historical Mindsets

It is a common error to apply modern thinking to historical figures to critique their ethics, morals, or decisions. While it is important to draw modern lessons from a battlefield study, this is different than applying modern values. There is value in drawing out the differences between past and present modes of thinking, though ideally without judgement. This keeps the focus of the activity on teaching tactical lessons, rather than moralizing.

PRACTICAL EXAMPLE 1

Gettysburg National Military Park allows exploration of the three-day battle in 1863, but is too large and too complex for most simple battlefield studies. Instead, we can focus on a small portion of the overall battle, and use it to draw out lessons.

This battlefield study focuses on a single, tactically complex location during the Battle of Gettysburg: Devil's Den, on the Union left flank. While participants may be given time to explore the wider battlefield for general interest, this study is deliberately narrow in scope to ensure a focused and effective lesson. They should be given a simple handout that describes the context of the overall battle, and the specifics of what occurred in the Devil's Den.

- **Aim:** To examine how complex terrain influences tactical decision-making at the company level.
- **Scope:** The actions of a single Confederate infantry company during the assault on Devil's Den on 2 July 1863.
- **Training Objectives:** To analyze how irregular terrain affects movement, visibility, and unit cohesion, explore the challenges of command and control during an assault in broken ground, and assess leadership decisions made under conditions of friction, uncertainty, and fragmented contact.

Structure of the Study:

The group is led on foot across Devil's Den, pausing at key locations to discuss terrain features and how they impacted historical events. The instructor provides a brief historical overview at the start, including the relevant portion of Hood's Division's assault and the Union defense by elements such as the 124th New York. Participants should receive a short handout beforehand, including a map, a timeline of key events, and a summary of the forces involved.

The instructor guides the discussion by asking questions at relevant terrain features (e.g., covered approaches, observation points, likely rally points). Participants are asked to consider the ground from both attacker and defender perspectives.

For more senior groups, the instructor may assign roles in advance — for example, a participant may be asked to take on the role of Colonel Evander Law (commanding the assault) or Colonel Augustus Van Horne Ellis (defending Devil's Den). At key points during the walk, the instructor can then ask them to explain the decisions made by "their" commander, based on the historical context and their own interpretation of the situation.

Sample Discussion Questions:

Terrain and Tactics
- How would the terrain have affected your ability to maintain momentum during an assault?
- Where are the natural chokepoints or dead ground? Could they have been exploited?
- How would you coordinate supporting fires in this terrain? Could you even see your own flanking elements?

Command and Control
- What difficulties would a company commander face in issuing and adjusting orders here?
- How might confusion or terrain-induced isolation have changed the course of the engagement?
- Would you keep reserves close or farther back in this kind of environment?

Leadership and Decision-Making
- Why might a commander choose to commit his troops piecemeal in this ground, rather than massing them?
- What risks would a commander be willing to accept in order to seize this feature quickly?
- If you were on the receiving end of this assault, how would you reinforce or hold the position?

Role-Based Prompts (if using historical assignments)
- "Colonel Law, what did you believe the intent of your attack was at this point in the battle?"
- "Colonel Ellis, what was your assessment of the terrain's defensibility, and how did that affect your deployment?"

- "How did your understanding of the larger battle shape your immediate tactical decisions here?"

Duration:

This study can be completed in 60–90 minutes, depending on walking speed and the depth of discussion. It is best suited to junior leaders with some experience in terrain analysis, though it can be adapted for more senior personnel by deepening the role-play element and placing greater emphasis on command perspective and operational context.

PRACTICAL EXAMPLE 2

This virtual battlefield study uses satellite imagery and terrain tools such as Google Maps and Google Earth to examine a discrete element of the Battle of Kapyong, fought from 22–25 April 1951 during the Korean War. While the full battle involved several Commonwealth and UN units across a wide area, this study focuses specifically on the defensive actions of 'D' Company, 2nd Battalion, Princess Patricia's Canadian Light Infantry (2 PPCLI), holding a key hill feature against repeated assaults by Chinese forces.

- **Aim:** To study the role of terrain in a company-level defensive action and examine leadership decisions made under sustained pressure and isolation.
- **Scope:** The defense of Hill 677 by 'D' Company, 2 PPCLI, during the night of 24–25 April 1951.
- **Training Objectives:** To assess how terrain is used to shape a defensive position at company level, how a commander maintains control and combat power while isolated, and analyze decision-making under the pressure of encirclement and limited support.

Structure of the Study:

Participants access Google Maps or Google Earth, focusing on the Kapyong Valley, particularly Hill 677, located west of the Kapyong River. The instructor provides a short historical overview before the activity, along with a timeline and simplified map showing the disposition of 2 PPCLI and their likely fields of fire, as well as enemy avenues of approach.

Participants follow along as the instructor virtually "walks" the ground using satellite view and topographic overlays. Key terrain features — ridgelines, covered routes, fire positions, and dead ground — are identified and discussed. Photographs from the period or modern terrain comparison images may be included if available.

For more senior groups, instructors may assign roles such as playing Major Bernard Stone, commander of 'D' Company, or Chinese battalion commanders, and prompt the students to explain the rationale behind their decisions during key phases of the battle.

Sample Discussion Questions:

Terrain and Defense
- Why was Hill 677 selected as the main position to hold?
- What advantages and disadvantages does this terrain offer to defenders?
- How might you position platoons to mutually support each other on this terrain?

Command and Control
- How could a company commander maintain situational awareness without clear contact with battalion HQ?
- What would you do to preserve combat power after sustaining casualties and facing isolation?
- How does the terrain affect your ability to shift forces or resupply under fire?

Leadership Under Pressure
- What decisions had to be made once 'D' Company was surrounded and communications were intermittent?
- What leadership traits are necessary in this situation, and how did Major Stone demonstrate them?
- How would you maintain morale and cohesion under sustained assault and isolation?

Role-Based Prompts (if using historical assignments)
- "Major Stone, what was your intent when you chose your platoon dispositions on Hill 677?"
- "As the Chinese commander, how would you exploit the terrain to dislodge a dug-in enemy?"
- "What role did supporting fires play in your plan, and how did limited visibility affect their coordination?"

Duration:

This virtual study can be conducted in approximately 45–60 minutes, depending on the depth of discussion and number of participants. It can be conducted in a classroom setting with a shared screen or individually with guided materials. For best results, ensure participants have reviewed a short background reading on the battle and are familiar with basic principles of defense.

NOTES

NOTES

METHOD 4 - TACTICAL EXERCISES WITHOUT TROOPS (TEWTS)

DOES THIS METHOD TEACH:			
Tactical Acumen	Tactical Awareness	Speed of Decision-Making	Tactical Agility
✓	✗	✓	✗

A TEWT is a resource-efficient way to conduct otherwise complex training, often with equivalent outcomes for many of the participants when compared to a Field Training Exercise (FTX). Normally, a TEWT is conducted at higher levels than we're concerned with here, but we'll get to that. First, let's make sure we understand what a TEWT is.

A typical field exercise might see a leader receive orders, conduct a planning cycle and then execute the operation according to their plan. But other than during the execution portion of the exercise, most of the soldiers, vehicles and other physical resources that make up the leader's organization are not required. In some cases, soldiers "sit on their duffle bags" waiting for the planning cycle to be completed. By contrast, a TEWT uses only the resources needed to allow the leaders to practice their planning skills.

The basic requirements for a TEWT are simple:

- A set of orders from "higher" headquarters to act as the input,
- A specific output that the participants need to produce.

The input orders become the prompt for the start of a planning cycle. This cycle could be one conducted by a headquarters staff, or by an individual commander, depending on who and what is being exercised. The training objectives of the exercise will determine the correct input.

The second requirement, the desired output for the exercise, focuses on the training objective. For any given input, there are many different options. For example, a company commander might receive battalion-level orders as the exercise "input." The output could be a mission analysis brief, a written estimate, a verbal briefing on the key deductions in the estimate, a written set of orders, or the verbal delivery of orders.

Typically, TEWTs are conducted to exercise commanders and their staffs. For example, providing a Battalion Commander and staff with a detailed set of brigade orders can be used to launch a planning cycle that focuses on roles within the headquarters, the planning process, and the generation of orders and other staff products. This is important training to conduct but is only peripherally connected to teaching them tactics. A well-oiled headquarters increases the speed at which a commander can consider problems and issue orders, but this is in support of good tactics rather than an expression of it

TEWTs and Low-Level Tactics

In order for a TEWT to be effective at teaching tactics at lower levels, the focus must be less on teaching process and more about generating critical thought. To achieve this with a TEWT, the emphasis should be placed on discussion of why certain decisions have been made (i.e. the selection of a left flanking versus a right flanking during a planned attack), rather than on the resulting product.

The training output desired must be clear, but as described in the example above, there are many options. The simplest is that the student generate a set of written or oral orders. The emphasis on critical thought rather than on product is achieved through questioning, forcing the student to explain and defend his or her decisions.

When the TEWT is conducted using orders connected to terrain that is available to the participants, they can conduct a reconnaissance as they would during a field exercise (or operations). The commander's reconnaissance can also be conducted as a group activity, with various stops on the route used to generate discussion around relevant aspects of tactics. For example, student might want to observe the site of a potential enemy river crossing. This could spur discussion on enemy vs friendly amphibious capabilities, OPFOR river crossing drills/tactics, the positioning and use of Named Areas of Interest (NAI) and Target Areas of Interest (TAI), etc. The real training value of a TEWT could easily be derived from these discussions more than from the students' planning activities. In these cases, the students' planning simply serves to make the discussions well informed.

This form of a TEWT is also useful for exercising junior leaders whose planning process in their assigned roles is very limited. Involvement in the discussion during the reconnaissance might be the only output required of them. A TEWT is an excellent opportunity to allow junior commanders to consider problems that would normally be "above their pay grade," as the resource cost is limited. As an example, platoon commanders can be given a company level problem that they are required to consider as if they were the company commander. This is challenging training for them which builds their tactical knowledge and encourages the future use of initiative.

Typically, we exercise ourselves from a "Blue Force" perspective, with the OPFOR acting as a supporting element to ensure that training objectives are met. But this does not have to be the case. It can be very illuminating to conduct a TEWT from the OPFOR perspective, using OPFOR ORBATs and, most importantly, OPFOR doctrine. A student could be tasked to plan the OPFOR response to a previously "solved" Blue Force problem, essentially thinking about how to defeat ourselves. This can make for a very effective two-part series of TEWTs.

Another area where TEWTs have tremendous value is in training for operations in built up areas. Many of the training sites used for field exercises focused on these types of operations do not fully represent the full complexity of dense urban terrain. Understanding the impact of this sort of terrain can be achieved through a TEWT held on ground that could not be used for a field exercise, such as the center of a major city. Students can explore and plan for operations in this terrain, getting nearly all of the benefit from the TEWT that they would from conducting a live exercise.

Finally, historical scenarios can also be used to conduct TEWTs, either using the original orders (sometimes available in online archives) or by using summaries prepared for the purpose. In these instances, the focus should be even less on product and more on critical thinking about enduring military concepts such as the use/importance of reserve, surprise and deception, etc. This approach blurs the line between this method and the battlefield study, but as long as the training has a clear aim and this is carefully maintained, this doesn't matter.

A TEWT differs from a Field Problem in one key way: the complexity of the input. The focus of a Field Problem is on a simple, often quickly devised, problem. A TEWT, on the other hand, uses formal products such as written orders to provide information to the participants. This allows for the consideration of much more complex problems, and also helps to familiarize the participants with the process of quickly digesting formal orders.

How Does This Method Teach the Different Components of Tactics?

Tactical Acumen

Recognizing the key elements of a tactical problem, the available and suitable means to overcome the problem, and the selection of the best of these options are all at the heart of the conduct of a TEWT. It is an excellent means to practice tactical acumen, especially when ample time is given to discussion of the thought process that led to a decision.

Tactical Awareness

Because the plans created during a TEWT are typically never executed, there are no opportunities for students to practice tactical awareness. In a limited sense, students may gain a better sense of how to read ground during the reconnaissance portion of a TEWT, this skill is often better trained with a simple Field Problem.

Speed of Decision-Making

It is very simple to impose time limits on a TEWT to allow students to practice rapid decision making. While this is a skill that often comes with experience, there is also an aspect of disciplined thinking that is required.

Tactical Agility

Although this method might aid students in recognizing what drills must form part of their plans (i.e. a crest drill during an advance that is identified through a map study), because the plans made are typically never executed, there is no opportunity to put drills into practice. It is not well suited to exercising tactical agility.

Critical Elements of the Method

Recycle Old Exercise Orders/Products

Rather than generating new orders to use for a TEWT, old products can be "recycled". The same orders can be used multiple times by assigning students to different sub-units, or by focusing on different portions of the operation.

Conduct the Exercise on Available Ground
Although it is possible to conduct a TEWT using any piece of ground for which you have maps, there is value in conducting the exercise using ground that the students can visit. There is training value in recognizing the difference between physical terrain and how it appears on the map, but more importantly, it helps them translate their mental conception of the operation into reality.

Use Computer Simulation to Test Solutions
One of the shortcomings of a TEWT is that there is no way to test the feasibility of the plans devised by the students. Feedback from an experienced mentor is useful, but this can easily devolve into the idea that "tactics are the purview of the senior officer present" should there be a disagreement over how successful a particular plan might be. Where there is simulation time available, plans can be tested against a thinking enemy, creating further opportunities for discussion and critical thinking.

Challenges when using the Method

Impose Strict Time Limits
Even when a TEWT is conducted intermittently over an extended period of time, it is important to impose appropriate time limits on the participants. Planning will often expand to take all of the time available unless strict discipline is imposed, with no appreciable improvement in the quality of the final decisions or products. Students must become accustomed to thinking quickly and thoroughly through a problem, so that even when time is short, the best possible decisions can be made.

Focus on Critical Thinking
We can easily find ourselves focusing on physical products as the desired output for a TEWT, but the real training value is in the thinking. Although thought is required to produce a set of orders, for example, critical thought is also generated through questioning and discussion. No matter what the selected end product of the activity is, ensure that adequate focus is maintained on *why* decisions are made, not only on *what* decisions are made.

PRACTICAL EXAMPLE 1

This TEWT is suitable for both company-level or platoon-level training, and is conducted by a company commander to train platoon commanders in the siting of company positions as part of a delay operation. It uses an old set of battalion orders for an exercise in the local training area, so the real terrain is accessible.

- **Aim:** To develop platoon commanders' ability to analyze terrain and site positions that support a delaying action within the framework of company and battalion intent.
- **Scope:** platoon commanders of an infantry company.
- **Training Objectives:** To identify suitable terrain for successive positions in support of a delay, evaluate the strengths and weaknesses of selected terrain from the defender's perspective and foster discussion around timing, spacing, and the trade-offs between defensibility and disengagement.

How the TEWT is Conducted:

Participants are issued battalion orders and given time to study the situation and the map. The company commander might also conduct a review of the relevant doctrine beforehand. Once the participants' preparation is complete, a list of the proposed company-sized delaying positions is provided by the company commander, which the group will visit, ideally in the sequence that they would be occupied.

At each delaying position, the platoon commanders are given 30 minutes to make a plan to site the company. One platoon commander is then selected to present his plan, which is discussed, along with other relevant doctrinal issues. The total time at each delaying position should be about one hour.

The platoon commander's brief must cover:

- Their understanding of why this position was selected.
- How platoons will be sited.
- What control measures are required.
- How the company will disengage.

The remainder of the group acts as peer evaluators and discussion participants. The real value of the TEWT is drawn from the discussion that follows each brief, with the company commander acting as a facilitator and ensuring the focus remains on key tactical considerations.

Sample Discussion Prompts:

- What makes this position tactically significant?
- Where is the natural limit of enemy observation and fire?
- How would you control platoon withdrawal from this position?
- Is this position worth holding, or is it simply a speed bump?
- Where would you place the reserve or blocking force?
- How does this position link to the next one in depth?

Notes for the Instructor:

- The emphasis is on **terrain-driven tactical decisions**, not on the perfection of orders.
- Allow participants to disagree with one another — this drives critical thinking.
- Keep briefs and discussion time-bound to preserve pace.
- Rotate roles to give each junior leader a chance to act as company commander.

The senior NCOs from each platoon could also participate in this activity, siting a single platoon at each delaying position. In this case, they would be the STA. This entire activity could also be conducted by a platoon commander, in which case senior NCOs would be the PTA.

PRACTICAL EXAMPLE 2

This example will work at both the company and the platoon level to train urban operations. The company commander can walk through a town with her platoon commanders to discuss how an intersection could be defended. It uses an old set of battalion orders for an exercise in the local training area, so the real terrain is accessible.

- **Aim:** To develop platoon commanders' ability to site a defensive position in an urban environment within the framework of company and battalion operations.
- **Scope:** platoon commanders of an infantry company.
- **Training Objectives:** To analyze terrain in an urban environment; to identify options for defending a specific type of urban position in support of a greater defensive position.

How the TEWT is Conducted:

Participants are issued battalion level orders and given time to study the situation prior to visiting the town, city or built-up area. The company commander might also conduct a review of the relevant doctrine beforehand. Once the participants' preparation is complete, the company commander explains how the battalion (or company) will defend and instruct the platoon commanders to consider how the company (or platoons) will be deployed.

At each position, the platoon commanders are given 30 minutes to site the company (or his platoon). One platoon commander is then selected to present his plan, which is discussed, along with other relevant doctrinal issues. The total time at each position should be about one hour.

The platoon commander's brief must cover:

- His understanding of why this position was selected.
- How platoons (companies) will be sited.
- What control measures are required.
- How the companies (platoons) will support each other.

The remainder of the group acts as peer evaluators and discussion participants. The real value of the TEWT is drawn from the discussion that follows each brief, with the company commander acting as a facilitator and ensuring the focus remains on key tactical considerations.

Sample Discussion Prompts:

- What makes this position tactically significant?
- Where is the natural limit of enemy observation and fire?
- How would you control withdrawal from this position?
- Where would you place the reserve?
- How does this position support other positions in this defense?

Notes for the Instructor:

- The emphasis is on terrain-driven tactical decisions, not on the perfection of orders.
- Allow participants to disagree with one another — this drives critical thinking.
- Keep briefs and discussion time-bound to preserve pace. (remember the special note above).
- Rotate roles to give each junior leader a chance to act as company commander.

The senior NCOs from each platoon could also participate in this activity, siting their sections or squads withing the platoon position. This would make them the STA. This entire activity could also be conducted by a platoon commander, in which case senior NCOs would become the PTA.

NOTES

NOTES

_____

NOTES

METHOD 5 - TACTICAL DECISION GAMES (TDG)

DOES THIS METHOD TEACH:			
Tactical Acumen	Tactical Awareness	Speed of Decision-Making	Tactical Agility
✓	✗	✓	✗

Popular in military magazines in the 1980s and 1990s, tactical decision games (TDGs) are a simple way to asynchronously exercise leaders and teach tactics. Many past issues of military magazines such as the United States Army publications *Armor* and *Infantry* are available online, and the tactical decision games in them remain relevant and can still be used for training. There has been a recent resurgence of TDGs, such as those that can be found in in the Canadian Army Journal and the associated website Line of Sight. The Royal Canadian Regiment has even launched a TDG competition. This trend is only likely to continue.

A tactical decision game is a short piece of text that describes a situation faced by a military element, often accompanied by a map or sketch. The size of the military element can be anything from a squad to a corps, or even non-standard units such as a Viking "horde." The student considers the problem posed by the situation described in the text and produces a response as directed. This could be a written Frag O, a sketch, or any other simple answer to the problem.

In their original form, the responses to tactical decision games were sent to the magazine, and answers published in subsequent issues. Used as a training tool within a unit, there are many possibilities to adapt this method. Problems could be posed in a regularly scheduled weekly or monthly meeting, and solutions discussed at the next one. They could also be posted on a unit website or SharePoint site. However they are distributed, they allow for students to consider and respond to the problem on their own time. They are ideally suited to ongoing professional development plans, and can be part of a series of lessons, and indeed an ongoing conversation, about tactics.

As with Field Problems, there is not necessarily a single right answer to the problem posed by the TDG. A clearly articulated justification for the decision made can form part of the required solution, or be an integral part of the discussion of solutions. The key result of a TDG is causing the particpant to think and communicate their thinking, rather than being a vehicle for finding the "correct" solution to a problem. Once participants understand this, they will become more willing to share creative solutions and, by doing so, learn.

There are three components that are required when creating a TDG:

- Situation
- Sketch, and
- Statement of the Problem.

Situation

A short description of what is going on, what forces are involved, and the context in which everything is occurring. This could be a mission statement, an extract from orders, or a simple description. While the situation and forces involved can be anything, and do not necessarily need to closely relate to the target audience of the problem, it is especially important that the context be given in a form that is appropriate to them. Untrained or junior soldiers should not be expected to parse the precise military language of a Concept of Operations (CONOPS) for meaning but could simply be given a plain language statement to explain what is going on.

Sketch

Practically all military problems benefit from a sketch. A sketch serves to create the habit of seeing problems (and their solutions) visually, which is a critical skill for military leaders.

Statement of the Problem

This is the deliverable for the student to complete. It should be a simple task and require careful consideration of the situation. An example would be to produce a radio Frag Order (FRAGO), a simple sketch of a scheme of maneuver, or a mission statement. A very focused problem might simply require the student to identify a main effort or determine trigger criteria for the use of a reserve. The statement of the problem must closely align with the learning objective of the TDG.

How Does This Method Teach the Different Components of Tactics?

Tactical Acumen
This method forces students to study a situation and determine the most appropriate actions to take, which is the consideration at the heart of tactical acumen. Discussion at the conclusion of the training activity will draw out the critical thinking that is a necessary component of tactical acumen.

Tactical Awareness
As with many of the methods in this section, this approach does not effectively teach tactical awareness since there is no "live" component.

Speed of Decision-Making
By imposing time limits, or more likely directing students to impose limits on themselves, rapid decision-making can be practised.

Tactical Agility
As with many of the methods in this section, this approach does not effectively teach tactical agility.

Critical Elements of the Method

Create a Library of Problems
Leaders can collect TDGs that they have devised for training, as well as those that they find online in magazines, etc., to create a library of options for future use. With just a few tweaks, an old problem can be given new life, focusing on a different aspect of tactics. Old problems can also be updated to fit within conflict scenarios playing out in the news, giving them a sense of relevance and immediacy that often sparks greater interest among participants.

Build a Culture of Tactical Thinking and Learning
The continual employment of TDGs as a training tool within a unit plays a part in creating a culture of tactical interest. This can be a regular part of the

unit's routine, such as a weekly "Tactics Tuesday" post. If the commander of a unit shows interest in discussing tactics, their subordinates will naturally do the same, thereby elevating everyone's tactical acumen.

Know the Learning Objective When Creating a Problem

It's easy to get lost in the details of a problem scenario, creating complexity that detracts from the intended lessons. Students can often focus on irrelevancies if the problem isn't written in a way to nudge them towards focusing on the issues that the leader desires.

Challenges When Using the Method

Too much focus on the "right answer" will stifle thinking. The majority of the real learning occurs through critical thinking and discussion, rather than reaching the same conclusion as the leader who devised the problem. While there are certainly some solutions that will be better than others, there must be room to explore the idea that all answers involve trade-offs and different perspectives on risk.

Incentivize Participation

Some students will naturally feel uncomfortable sharing their solutions in public. Leaders should consider how to incentivize all participation, rather than simply rewarding those who get the "right" answer. Even a sub-optimal solution provides training value, as critical thinking is necessary to determine and explain where it falls short. The simple act of sharing a solution should be praised. Recall the example above of The Scotty Plan (page 42).

Ensure That You Follow Up With Discussion

It isn't enough to circulate a problem to subordinates and collect their responses. Discussion is a key element of learning, where students are given the opportunity to explain and justify their responses, question the thinking of others, and grow their understanding of tactics.

PRACTICAL EXAMPLE 1

Frequently, a Mission Analysis (MA) is used as a catch phrase where commanders and subordinates blindly run through the sequence without fully appreciating *why* the process is critical to their decision-making. In this case, a rapid MA is forced upon the leader. The solution is binary, but the value of the training is in appreciating the *reason* for opting for one of the two options. As always, the focus of the acitvty is on developing the ability of the participants to think through the problem rather than simply arriving at a solution.

- **Aim:** To practise rapid Mission Analysis.
- **Scope:** Company level operations.
- **Training Objectives:** Appreciate the importance of what it means when a situation changes; reinforce the value of bold action; stress the value of initiative.

Situation

You are commanding a mechanized company in the vanguard of an advancing force. Your orders tell you that there is no enemy to be expected on this side of the river. Your commander two levels up intends to cross the river to force the enemy to fight a decisive engagement at the objective and is therefore pushing you and your commander to move quickly. Speed is paramount.

As you advance, a dispatch rider arrives with orders for you to seize the crossing immediately. You ack the order, do a quick mission analysis, and give your own orders on the move. As you approach the bridge you are ambushed. Although you successfully defeat your enemy, your unit suffers heavy casualties and is reduced to only 40% of its original strength.

You have no comms with your superior.

What do you do next?

Statement of the Problem

Your solution should be in the form of a simple verbal order. It must be SUCCINCT and CLEAR.

For the sake of simplicity we'll confine the solution to one of these two options:

- Seize the crossing, or
- Wait for your commander to arrive.

REMEMBER: Your solution must explain not only *what*, *where*, and *how* but <u>most importantly</u>, you must explain ***why***.

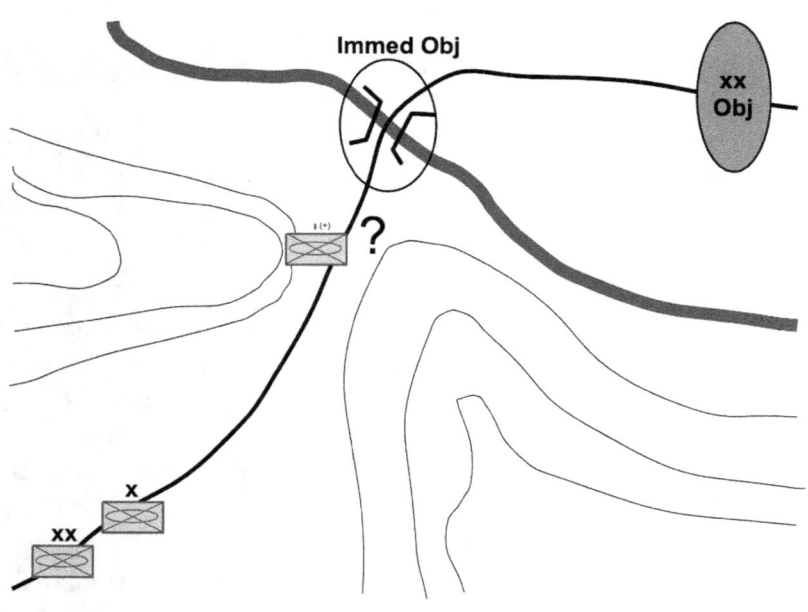

Practical Example 2

Commanders have to think and react on the move. In this case, the instructor (at any level) presents a scenario where a decision needs to be made quickly and acted upon just as quickly. This problem can be presented anywhere with minimal tools and preparations. It can also be used as a classroom presentation to demonstrate to all soldiers how important it is to have leaders who can think on their feet. The solution is purposely kept to three possibilities for the sake of simplicity.

- **Aim:** To practise rapid decision-making.
- **Scope:** Applicable at any level but most applicable at company level operations.
- **Training Objectives:** Practice quick actions; appreciate the importance of what it means when a situation changes; reinforce the value of bold action; stress the value of initiative.

Situation

You are commanding a mechanized unit (size is irrelevant) that is part of an advancing force. Your commander has been ordered to seize control of the crossroads at the town of Adorf. She has sent you forward as the Advance Guard. As you move to accomplish this, an enemy force only slightly smaller than the one you personally command appears on the left flank and threatens the mission from the hill on your side of the river.

Your commander, as smooth as glass, orders you to eliminate the threat. As you move to do so, the enemy withdraws across the river and outside your boundaries.

You have no comms with your superior.

What do you do next?

Statement of the Problem

What do you do? For the sake of simplicity we'll confine the solution to one of these three options:

- Chase the enemy across the river and destroy him,

- Seize the ground that the enemy was occupying, or
- Establish a flank guard on your side of the river.

REMEMBER: Your solution must explain not only *what*, *where*, and *how* but <u>most importantly</u>, you must explain ***why***.

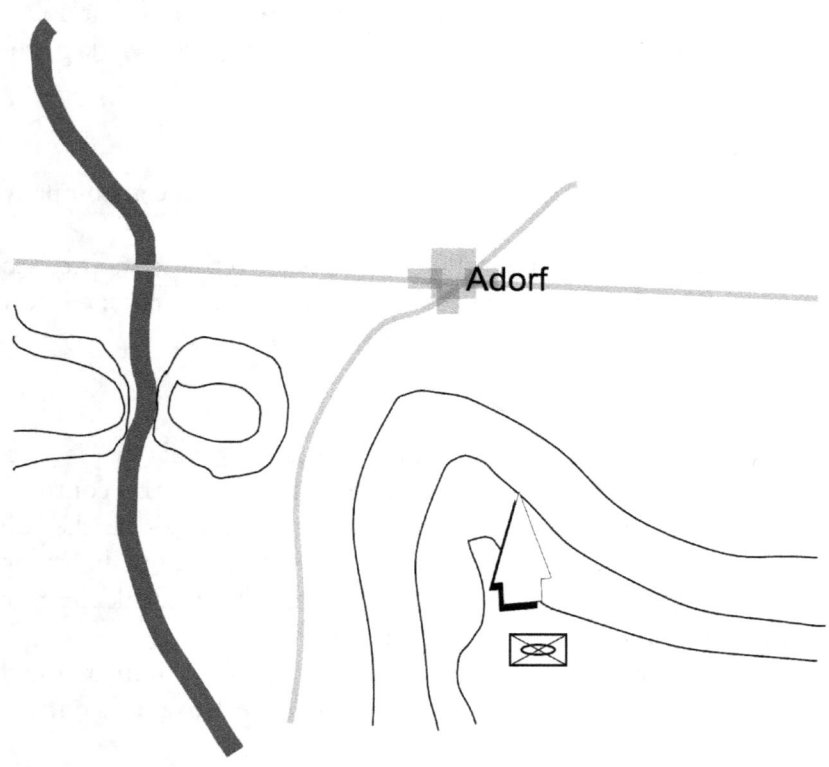

NOTES

NOTES

METHOD 6 - TACTICAL DISCUSSION

DOES THIS METHOD TEACH:			
Tactical Acumen	Tactical Awareness	Speed of Decision-Making	Tactical Agility
✓	✓	✓	✗

Tactical discussions are excellent training tools whose usefulness spans the entire spectrum of training. They can be short and simple or extended and complex. As a young leader, you might feel under-qualified to conduct such a discussion, especially if your subordinates are more experienced than you. But, as the leader conducting the discussion, you don't need to have all the answers. You just need to be able to manage the discussion, generate ideas from the group, and relate it to the relevant doctrine or principles. There doesn't need to be a final "answer" to the discussion for it to be useful.

To be effective, decide on one or two tactics that you want to teach and focus on them. As always, keep it simple. Since you are presenting a tactical problem, focus your troops' attention with a map or diagram. This can be a white board, drawn on the side of an armored fighting vehicle with chalk, a mud model, paper map or screen with a digital map displayed. Use the simplest tool that will accomplish the aim. It is possible to hold a discussion without a visual aid, but the problem must then be simple to envision. The key is to remain focused on the tactic you chose as the training objective.

Begin by stating the tactical problem in simple terms, avoiding any unnecessary detail. Do not purposely include information that will confuse or misdirect. This is a lesson, not a contest. Once the problem has been stated, open the floor to questions. If there are no questions, choose someone and ask for a suggestion on how to begin. Do not ask for a solution. The advantage of this sort of discussion is that it is like time-stop photography; you get to stop the clock whenever you want to ensure that everyone has situational awareness.

Once you have established a potential point of departure, ask the ever-important question: "So, what?" In other words, "What deductions can we draw from where we stand?" Each time a new deduction is drawn (preferably by one of the students), you should as the rhetorical question "So what?" again until no new deductions can be drawn. When there are no new deductions, it's time to move to the next fact. Initially, there may not be much to draw out, but you have now established the routine that you will repeat until the discussion has reached its conclusion. Now that everyone is more or less situationally aware, begin the cycle again. Pay attention to your training audience. If someone seems puzzled or confused, do not move on. Gently review where you are in the tactical scenario to ensure that you haven't lost anyone, then proceed.

As before, ask someone to suggest the next tactical move. Again, if there are no volunteers, choose someone and work with their suggestion until you achieve a workable position. Remember that as long as you are proceeding in a workable manner, the solution may not be the one you imagined. That is fine, so long as it leads to a viable method. Never dismiss an offered suggestion. It is easy to lose a subordinate by dismissing them or implying that their tactic was wrong. Instead, explain that such a move might work but, in this instance, you were thinking that you wanted to follow a different course of action.

Using this process, you can usually draw out the lesson that you had selected to teach. Depending on your own experience both as a tactician and as an instructor, it may be slow going initially. Don't be put off. If you keep it simple and you follow the steps, this form of discussion can be highly effective at teaching low-level tactics.

How Does This Method Teach the Different Components of Tactics?

Tactical Acumen

Discussions encourage participants to brainstorm and evaluate multiple solutions to a tactical problem. This process develops their ability to recognize various approaches and potential solutions to complex scenarios.

Tactical Awareness
By repeatedly assessing the current situation and asking, "So what?" participants improve their ability to understand the implications of tactical changes and make informed decisions in real-time.

Speed of Decision-Making
Frequent participation in tactical discussions develops confidence and clarity, enabling participants to make quicker, more effective decisions under pressure in real-world scenarios.

Tactical Agility
Without the physical deployment of troops, there is no opportunity to train tactical agility.

Critical Elements of the Method

Use the expertise available to you
You may well have experienced soldiers and NCOs in the group. Use their skill and experience to keep the discussion going, to do reality checks and to reinforce successful suggestions. A very junior leader may feel intimidated by the idea of leading a discussion with more experienced subordinates. This is the wrong way to think about – experienced subordinates will make the discussion easier, not harder, to conduct.

Link the Discussion to Drills
When a suggestion leads to a known drill or procedure, reinforce the suggestion by reviewing the drill or procedure to ensure that everyone can make the connection between drills, procedures, and successful tactics.

Stay positive. Even unhelpful suggestions have merit
Never dismiss or insult anyone who offers an unworkable suggestion. Turn it around and demonstrate how doing what they suggest might work but is likely to lead to more difficulty. It's likely that the poor suggestion stems from a misunderstanding of relative risks or a misapplication of a drill – these are great teaching points.

Challenges When Using this Method

Maintain Focus

People may attempt to lead the discussion in unhelpful directions, such as focusing on personal expertise or unrelated topics. The facilitator must keep the group focused on the lesson objectives.

Uneven Participation

Some participants may dominate the discussion while others remain quiet. Ensure everyone contributes by directly involving quieter members and setting expectations for balanced participation.

Confusion

Participants may become overwhelmed if the problem is overly complex or poorly explained. Simplify the problem, use visual aids, and review progress frequently to ensure understanding.

Practical Example

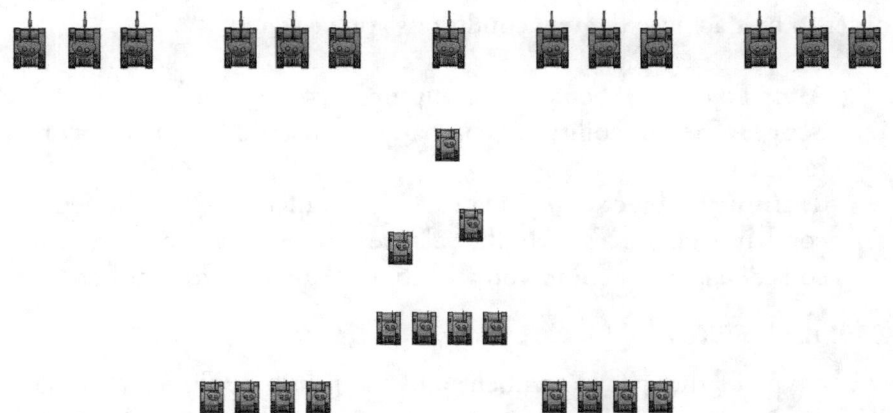

In this example, you have recently completed the same problem as part of an officers' study session and the Battalion Commander has made two points that were both new to you and which he wanted passed to all soldiers:

- Do not destroy command relationships – build them;
- Balance is usually a desirable end, but not at the cost of combat power.

The task given to start the discussion is for participants to array the forces shown, a tank company and a mechanized infantry company, as they would be grouped and deployed for an advance to contact in order to achieve maximum flexibility and combat power. This could be achieved with models on the ground, by having students draw their organization on a piece of paper, or many other methods.

Situation

You are commanding a company combat team consisting of a tank company and a mechanized infantry company. Using the diagram below, explain to your students that in the advance, you wish to ensure two things: maximum combat power and a balanced footprint on the move. Give them the diagram (or put it on a chalk board or something similar). Begin with some ideas like, tanks leading with one platoon in the rear as a reserve to get them started, then give them 30 mins to consider how they would lay out the combat team in order to best achieve the two conditions you've stated.

- **Aim:** To discuss the idea of grouping of combat forces.
- **Scope:** The variability of combat power at the company combat team level.
- **Training Objectives:** Practice combining two components of combat arms forces to build balance; how to build combat power; considerations of command and control of mixed formations.

Note for the Instructor:

This is a type of thought experiment and the possible solutions are nearly endless. Present the problem and give everyone a time limit. Use the expertise of your NCOs. Do not worry if the numbers are not representative of what a tank company or infantry company in your formation look like. You may want to begin with a very brief reminder of what constitutes combat power as well as your army's guidelines and doctrine regarding the grouping of forces and how it impacts command and control.

Don't tell everyone unless asked, but be ready to answer:

- Potential enemy.
- Location of enemy.
- Type of ground
- Probable mission

The reason why you do not give this info at the start is to see how many will begin asking and then bring this up as a discussion point to emphasize the kinds of factors that can have big influences upon both groupings and tactical configurations.

Do NOT deny potential answers simply because they are not doctrinal. Allow open discussion by the entire group to draw out potential advantages and risks.

NOTES

NOTES

METHOD 7 - TABLE TOP EXERCISES

DOES THIS METHOD TEACH:			
Tactical Acumen	Tactical Awareness	Speed of Decision-Making	Tactical Agility
✓	✓	✓	✗

Table Top Exercises (TTXs) are excellent training tools whose usefulness spans the entire spectrum of training. Many professional militaries gather senior commanders on a regular basis to conduct large operational/strategic level TTXs. Here we'll focus on using a TTX to teach low-level tactical lessons.

TTXs look a lot like other similar military activities that are often conducted by staff and commanders as part of the planning process. These are COA wargames and Rehearsal of Concept (ROC) drills. Although superficially similar, these are different enough from TTXs that it is useful to draw out the differences.

COA Wargames focus on evaluating multiple courses of action during planning, identifying potential risks, and ensuring feasibility. They are decision-making tools rather than training exercises.

ROC Drills are designed to confirm understanding of an approved plan, aligning all participants with the intent and sequence of actions. They are more about coordination than exploration.

TTXs, by contrast, are interactive training events where participants play different roles, including friendly forces, enemy forces, and other sub-elements. The focus is on executing a mission, solving problems, and learning from the outcomes through structured discussion.

In a TTX, students play different sub-elements of a tactical scenario, including friendly forces and possibly the enemy. The exercise typically incorporates the execution of a task that has been issued from a higher command level and for which participants have developed a plan. Each participant or group is assigned a clear role, ensuring that everyone knows their responsibilities and objectives within the scenario.

The Instructor's Role

The instructor acts as the umpire, managing the flow of the exercise, adjudicating outcomes, and sometimes playing the role of the enemy. The instructor introduces elements to the gameplay that challenge the students and highlight key tactical lessons. This might include unexpected events, changes in enemy behavior, or environmental complications. The key to these injects is that they are always meant to teach or underline a key lesson, not to baffle the students.

Gameplay and Adjudication

The exercise progresses as participants make decisions and execute their plans. Combat and other actions can be adjudicated using tables or the instructor's experience. The process of adjudication is less important than the discussion it generates. Gameplay can be stopped at any time to address a problem, analyze a decision, or discuss the implications of an action.

The instructor should control the gameplay by imposing "turns" on the game, but these can be of varying length. It works best if every player has the opportunity to execute one action per turn, and that the enemy is then given the opportunity to react. The play can also be reversed, with the players reacting to the enemy. Turns can also be based on a set period of time, with actions played out on the map based on what is possible in that period of time, but this requires greater adjudication by the umpire and detailed recordkeeping to make it work. Games often flow more easily if the umpire speeds up or slows down time, and therefore what happens within a turn, based on what is occurring in the game.

The emphasis is on learning through interaction, collaboration, and reflection. This iterative process helps participants refine their tactical skills and understanding.

How Does This Method Teach the Different Components of Tactics?

Tactical Acumen

By assigning participants roles and responsibilities, TTXs force them to think critically about the scenario and their actions. The collaborative nature of the exercise helps them see problems and solutions from multiple perspectives, broadening their tactical understanding.

Tactical Awareness

As participants respond to dynamic challenges introduced by the instructor, they develop a better understanding of how to assess the current situation and adapt their actions accordingly. This reinforces the ability to remain situationally aware in real-world operations.

Speed of Decision-Making

TTXs simulate real-time decision-making under pressure. By repeatedly practising the cycle of assessment, decision, and action, participants improve their ability to make timely and appropriate choices.

Tactical Agility

Without the deployment of physical forces, tactical agility is not trained by this method.

Critical Elements of the Method

Clear Roles and Objectives

Assign clear roles to all participants, ensuring each person understands their position within the scenario, their authority, and limitations. This clarity helps participants see how their individual actions contribute to the overall mission and supports meaningful decision-making.

Create Dynamic Challenges

The instructor should introduce unexpected events, friction points, or enemy actions that disrupt the participants' plans. These injects force critical thinking and encourage adaptive, flexible responses to emerging problems — just as in real operations.

Stop for Discussion

Use the flexibility of a tabletop exercise to pause the scenario when a teaching point arises or a significant decision is made. This allows participants to reflect on the choices they've made, explore alternatives, and consolidate learning in real time.

Challenges When Using This Method

Avoid Role Confusion

Participants may not fully understand their assigned roles, leading to overlapping responsibilities, confusion, or miscommunication during key decision points. The instructor must clearly outline each role at the beginning and reinforce responsibilities throughout the exercise as needed.

Avoid Overemphasis on Adjudication

Focusing too much on the mechanics of adjudicating outcomes (e.g., consulting charts, rules, or dice rolls) can derail the flow of discussion. The adjudication process should support — not overshadow — the learning objectives, and a well-prepared instructor can keep it efficient and unobtrusive.

Maintain Player Engagement

Participants may lose interest if the scenario progresses too slowly, becomes overly complicated, or lacks meaningful choices. The instructor must monitor group energy, manage the pace, and adjust complexity on the fly to maintain interest and ensure productive engagement.

Note on Training Design

The most common error when using computer aided simulation for training is to choose the simulation, then design the training. This is backward. Design your training based on the logical sequence of aim, scope and training objectives, etc., THEN decide on computer simulation and what type (see Annex B).

Practical Example

In this example, we're practising an advance to contact over varied terrain, using a force comprised of a tank platoon and an infantry platoon combined. This could be in the context of a larger, company combat team level scenario, but that isn't necessary for this activity. The training could be led by either platoon commander, and within the training scenario, it will also have to be clear which commander is leading the combined force at the time.

The PTA are all the tank crew commanders and the section or squad leaders from the infantry platoons. The STA could be other members of the platoons who would act as observers.

You have organized the training to achieve the following goals:

- **Aim:** To practise an advance to contact.
- **Scope:** Combined tank and infantry platoons.
- **Training Objectives:** Practise combining two components of combat arms forces to build balance; considering how to build combat power; considerations of command and control of mixed formations; drills during an advance to contact.

The TTX requires some representation of the terrain and your vehicles in order to be played. This could be a map (ideally a very large one, real or virtual), or a terrain model. The platoons would receive truncated orders that focus on the Situation, Mission and Execution. The PTA must be given time to consider the orders and make their own plans. The exercise can begin with the platoons in an assembly area, or at the line of departure.

As the instructor moves the exercise forward in time, new tactical problems are introduced that require reactions from the PTA. Examples of issues that can be injected into the training include:

The PTA should be asked to not only describe their actions, but to conduct the routine communications that are needed during operations – sending situation reports, contact reports, crew briefings, etc. The instructor should take advantage of their ability to stop or slow the action in order to ensure that issues are fully explored, and the lessons clearly learned. Overall, the exercise should give the PTA a good sense of how an advance to contact might unfold in real life and what the friction points might be.

This planning phase should include coordination between the tank and infantry elements, with particular attention paid to movement timings, fields of fire, and communication protocols. PTA should be encouraged to anticipate how terrain may affect vehicle movement and dismounted manoeuvre, and to consider contingencies for vehicle breakdowns, loss of comms, or unexpected enemy contact.

By the end of the session, participants should have a clearer understanding of the complexities involved in leading mixed-arms formations and be better prepared to make rapid decisions under pressure.

As the instructor, take the time to walk slowly through any issues that come up that seem unfamiliar or confused, and equally move quicky through the aspects that are already well understood. Your guidance will directly shape the learning experience.

NOTES

PART THREE
FIELD TRAINING

INTRODUCTION

Let's consider how to use your time in the field as an opportunity to teach tactics to your subordinates. Too often, young leaders fail to seize upon opportunities to train subordinates because they think that once they're in the field, there isn't time to teach tactics on their own.

This is a mistake.

There are opportunities to train and teach tactics in the field that are not available in garrison. It's only during field training that all the moving parts of a military organization are put together and forced to function as they would during operations. While there is often a desire to simplify aspects of field training, such as allowing logistic organizations to conduct their work in a non-tactical manner, or relying on resources that would not be available during operations, this approach can mask issues that later cost lives.

The key to effective field training is clearly identifying what and who are to be trained, and not allowing the organization to become distracted from these goals. Training design is always important, but especially so in the field due to the enormous resources that are often required.

It is relatively rare for a junior leader to be able to take their command into the field independent of a larger exercise. The sense of merely being a player trapped within a much larger scheme falsely leads some leaders to believe that they have no role to play in training tactics once they are deployed. Nothing could be further from the truth, though conducting training within a larger exercise takes some imagination and careful planning.

Fundamentally, there are two situations that you might find yourself in:

- You may be conducting field training where you are not in control. You are part of a larger scale exercise or series of training events.
- You may in the field with your troops for the express purpose of training them and be in complete control of all aspects of the training.

Each of these solutions requires a different approach.

Training When You're Not in Charge

If you are conducting training within a larger exercise, you will need to do so while meeting your directed timings, goals, movements, etc. This can be a challenge, but don't give up. You can use whatever opportunities you are given to enhance the time you are spending in the field with your soldiers. If you are not part of the PTA, consider it a blessing, because if you are, it is highly unlikely that you will have the time or the opportunity to make time to teach your troops something new outside of the scope of what is planned.

Nonetheless, you need to be alert to possibilities that may arise. Here are the most likely methods open to you:

- Tactical Discussions (covered in Part 1 - Simple Training)
- Tactical Exercises Without Troops (covered in Part 1 - Simple Training)
- Stand Training (covered below)

The three categories above are not mutually exclusive, and once you become adept at teaching tactics, many combinations are possible.

Training When You Are in Charge

Training under these circumstances offers almost complete freedom, limited only by time and resources. The time you are going to spend in the field with your soldiers will be controlled by you, so plan it well and use it wisely.

Resource issues aside, you are able to select from the full suite of training methods. This includes all those covered in Part 1, but also the three methods described in Part 2:

- Stand Training (covered below)
- Force on Force (covered below)
- Live Fire (covered below)

Let's look at each of these methods in turn.

METHOD 8 - STAND TRAINING

DOES THIS METHOD TEACH:			
Tactical Acumen	Tactical Awareness	Speed of Decision-Making	Tactical Agility
✓	✓	✓	✓

Stand Training has gone out of fashion. This may be due to leaders not understanding how to create challenging stands that are both exciting and educational, though it remains a very efficient and effective way to teach tactics in the field.

At the most basic level, a "stand" is a location where a group of soldiers will conduct a single task as part of training. Multiple stands might be chained together, with groups moving through them over the course of a training day. This is often done if a larger group has to achieve some set of battle task standards. Oftentimes, stand training can seem boring – the task to be completed at each stand is relatively simple. That might be to conduct first aid, dry weapon handling, range estimation, or similar skills. But stand training can be much more than that.

It is possible to identify the discrete skills or drills that an element, say a section, needs to be able to perform to execute some larger task. For example, in order to conduct a dismounted raid, a platoon must be able to conduct cross-country movement to the objective, conduct counter-ambush drills en route as needed, conduct an attack by fire and an assault, exploit a sensitive site, handle prisoners, treat wounded, etc. In order to train the platoon to conduct a raid, you could take them to the field and conduct multiple raids.

But in doing so, it is likely that every section within the platoon will not get the opportunity to practice every necessary sub-task. It is also likely that you will not be able to focus attention on each of these sub-tasks, ensuring that they are all executed to a high standard. This is where stand training could be used. By rotating sections through a series of stands, each of which is focused on one sub-task, you can ensure that every section will practice every part of a raid. Then, you can put all the parts together and conduct the raid as a platoon, confident that every section is capable in all of the roles. After thorough stand training, you will need fewer iterations of the platoon-level raid to become effective at it.

What makes stand training particularly powerful is the ability to focus your training time and effort. In the example above, it would also allow you to conduct an AAR with each section after they had performed their task, and then allow them to immediately conduct the same sub-task again. This method – train, AAR, train – produces greater results with fewer resources and in a shorter period of time than conducting a complex series of actions followed by an AAR of the previous day or days of work. Stand training, which by its nature focuses on a narrow window of time, fits naturally with this model.

Stand training can also be very imaginative. A classic example is the "bait and switch" task, where a section is told that they are to move down a path or road to conduct one task, but along the way are ambushed and have to extract themselves by executing the correct drill. Simply by telling the section that "they're tactical once they step off," you've given them all the information they should need to be paying attention to their surroundings. Your subordinates will come to expect the unexpected and will be prepared to react to whatever is thrown at them, after a few instances of using this technique. But whatever happens during the first iteration of the surprise task, always give them the opportunity to AAR it and to then try again. This ensures that positive lessons are learned.

This style of stand training works well with mounted troops as well. Reconnaissance patrols or even individual tanks can move between stands, conducting critical sub-tasks that are brought together in later training. If conducted by higher headquarters, it could even be platoons or companies moving between stands. Cross-country navigation can be covered off simply by having the elements navigate between stands. The possibilities are nearly unlimited once you recognize the advantages that using stand training can provide.

How Does This Method Teach the Different Components of Tactics?

Tactical Acumen

This is an excellent method for teaching tactical acumen, and can be fine-tuned to ensure that you are teaching a particular aspect of this aspect of tactics as well. By carefully selecting a suite of skills or drills to focus on at a number of different stands, you can easily prepare your subordinates for a particular mission or situation.

Tactical Awareness

While it might seem that stand training is too "canned' to train tactical awareness, this is not true. There are many ways to introduce complications or surprises that force your soldiers to be aware of what is going on around them. These don't always have to be "bait and switch" type tasks – you can also create open ended stands where leaders have to decide the right course of action in an ambiguous situation.

Speed of Decision-Making

This is trained easily with this method, and in fact it can be used to put subordinate leaders into situations where they need to make rapid decisions over and over again. While in a regular field exercise there are often long lulls of inactivity – with this method the focus is on the "exciting parts."

Tactical Agility

This is a key element of tactics that is easily trained with this method. Not only are subordinate leaders can be made to react to the situation they are facing, but they also have the opportunity to rapidly AAR their actions and try it again.

Critical Elements of the Method

Carefully Plan your Content

Carefully breaking down a tactical task that you want to train into its component parts is necessary in order to conduct effective stand training. Think through the "building blocks" of the bigger task in order to make sure that you are giving subordinates experience with all the issues they may face.

Don't Overuse the "Bait and Switch"

Surprising subordinates with an unexpected task, such as an ambush that they have to fight their way out of, is a great method to maintain situational awareness and interest in the training. But if you overuse this method, trainees will be looking over their shoulder all the time and won't focus on the training you are trying to achieve.

After Action Review

Ensuring that groups have the opportunity to conduct AARs in the immediate aftermath of conducting a task, and that they then get the opportunity to put what they've learned to use, is of great importance. Not only does this empower your junior leaders to identify their own errors, but it also allows them to correct them nearly in real time. This increases the trust and confidence of their subordinates and gives the leaders opportunities to grow. A stand at which there is no AAR conducted is a missed opportunity.

Challenges When Using this Method

Keep it Interesting

Narrowly devised tasks that really drill down on a discrete skill can be a useful way to make sure that you are training vital elements of a larger task, but too many of this type of stand can be boring. Mix the content of your stand training to keep your soldiers interested – some stands might be simple and straightforward, while others may require quick thinking and rapid reactions.

Crosstalk

Too much cross talk between groups moving between stands can lessen the training value of activities that rely on surprise. You can avoid this by controlling the movement of subordinate elements, employing waiting areas for stands where you want to control what trainees can see, and by changing the disposition of the enemy, etc., between rotations.

Training the Trainers

This method uses a lot of leaders to conduct the training, at least one per stand. You will have to draw on leaders from across your organization, and potentially from outside of it, to make sure that your key leaders are themselves getting trained rather than training others. Within a company, for example, you might have to run a skeletal command post to command the exercise while all available leaders from your headquarters are running stands – including your sergeant major, company quartermaster, etc.

PRACTICAL EXAMPLE

In this example, you are a tank platoon commander preparing for a larger exercise where the battalion will be conducting offensive operations. You have been given one training day to use for your own purposes before the company starts to exercise together. You might decide to organize your training as follows:

- **Aim:** To prepare the tank platoon for offensive operations
- **Scope:** Platoon operations within a company group context.
- **Training Objectives**: Practising drills **on the advance (blind corner, short defile, etc),**

In order to achieve this training within a single day and to ensure that every crew practices each drill more than once, you set up four stands, each a few kilometers apart on terrain that is suitable for the drill that is being practised. Your stands are as follows:

- Blind Corner Drill
- Short Defile Drill
- Crest Drill
- Contact Drill

In order to achieve this training, you enlist assistance from the company staff to run stands. At each stand, the crew is given the chance to execute the drill, which is then AAR'd and practised again. After an agreed upon amount of time, vehicles proceed to the next stand, advancing through them in a circular fashion. By the end of the training, every crew has had time to practice, assess and practice each drill again, which would likely not have occurred if the platoon had simply conducted a practice advance to contact.

Of course, the four stand activities above are just examples, and many other activities could be conducted that would be equally valuable. The essential thing to remember is that the commander designing the training should be deliberate about what needs to be practised, to create a foundation of success for future activities.

NOTES

NOTES

METHOD 9 - FORCE-ON-FORCE

DOES THIS METHOD TEACH:			
Tactical Acumen	Tactical Awareness	Speed of Decision-Making	Tactical Agility
✓	✓	✓	✓

Force-on-force training is when the two opposing sides face each other in field training, with neither side being a "canned" enemy for the other. It is free play, within a controlled environment. This type of training always excites soldiers, and rightly so. It can be challenging, educational, competitive and fun — all at the same time. But it is also a method which does not always lend itself to teaching tactics.

There are typically two reasons why soldiers get excited about force-on-force training. Firstly, they want to "test" themselves against their peers — whether for bragging rights or to get a sense of where they fit in the pecking order. And secondly, they see force-on-force training as freeing them from the restraints found in a typical exercise, allowing them to finally use their creativity and know-how to "win." Both of these desires can lead to poor training, if not properly managed.

There is nothing wrong with training soldiers to want to "win." In fact, soldiers who don't have an innate desire to come out on top are probably not very good soldiers. But if left unchecked, this desire to win can blind soldiers to the training objectives of an activity. "Winning at all costs" can easily lead to gaming the system, ignoring good tactics in favor of shortcuts, or otherwise undercutting the training objectives. There is a lot of learning to be had in losing, and in fact there are often more lessons to be learned from defeat than from victory.

If the two forces are expected to engage each other in combat, there needs to be method of arbitration in place. This might be some kind of weapons effect simulator or simmunition. But if neither of these training tools are available, the use of umpires is an old school solution that can work very well. Umpires need to be senior staff whose tactical opinion carries weight, so that they can rule on the outcome of an engagement as it happens. They also need to be well versed in the training objectives, so that they don't derail the training by accident through their rulings. This doesn't mean that one side can't "lose" — or even lose very quickly. If that happens, the key is to quickly conduct an AAR of what happened and use the remaining time to reset and conduct the training again. This way, the lessons are captured and hopefully learned as well.

You might also consider constructing force-on-force training that does not require any form of adjudication. An example of this would be tasking reconnaissance patrols to map a defensive position without being spotted. The patrols are judged on the accuracy of the information they collect. The defensive element is judged on its ability to spot the reconnaissance patrols and report them to higher. While this kind of exercise could be conducted with some form of simulation, it doesn't suffer much without it.

The key to good force-on-force training is placing guardrails around the actions of the two forces to ensure that the activity achieves the training objective. This can easily be done through giving explicit orders with all of the typical constraints and restraints like timings, boundaries and control measures. This allows the opposing commanders to exercise free will and good tactics, while also ensuring that the training objectives are met.

How Does This Method Teach the Different Components of Tactics?

Tactical Acumen
This kind of exercise often creates a very fluid exercise, practising tactical acumen to the fullest extent. The looser the constraints on the participants, the more options they have to choose from when making their plans.

Tactical Awareness
Without scripted opportunities to ensure players have awareness of the situation around them, the need for the tactical awareness of participants is paramount.

Speed of Decision-Making

The fluidity described above also gives a marked tactical advantage to the leaders who are able to make good, rapid decisions.

Tactical Agility

Along with speed of decision-making, this method favors those elements who are agile in the field, able to rapidly respond to direction and the changing situation.

Critical Elements of the Method

Design Backward from Training Objectives

Force-on-force events must be designed backward from clearly articulated tactical training goals (e.g., coordination of platoon maneuvers under pressure, use of terrain, or decentralized command execution). Failing to do so, the activity may default to a glorified game of capture the flag with rifles. The training audience must know what is being evaluated and why.

Conduct Joint AARs

Although each side must conduct an AAR of its own activities and process, a joint AAR is also an highly useful method that allows for an exchange of perspectives. A skillful moderator can draw out of the participants information about what they knew, when they knew it, and what they decided to do about it in ways that highlight the interplay between two military forces.

As with any AAR, the key is to adhere to the major steps, while not constraining participation. Be careful not to turn the AAR into a validation of one side at the expense of the other (see Annex C). Be balanced. Remind everyone what the goal of each side was and how closely each side came to achieving that goal. It is not a travelogue, so do not fall into the trap of going back over everything that happened. Each side certainly did some things well, just as each side left room for improvement. Use these highs and lows as focal points to encourage discussion on what could be improved and what needs to be reinforced.

Use of Injects and Friction

Planned injects (e.g., comms failure, casualty evacuation, change of mission) add realistic friction. This reinforces decision-making under uncertainty and prevents "perfect" engagements that don't reflect combat's messiness. It also allows you to shape the exercise to ensure that the training goals are met.

Challenges When Using this Method

Maintaining Control

This method has the potential to spiral out of control and not achieve the desired training aims if the two forces are not properly constrained. Careful design of the exercise, and thoroughly briefing your subordinates (even if they are not aware of all the details of the exercise) can mitigate this problem. As exciting and useful as this training can be, control is essential. You will need trusted Observer/Controllers who are senior enough to step in if the training is going off course and who are experienced enough to decide when and where an intervention *might* be required.

"Hollywood Tactics" and Unrealistic Play

Without tight guidance, troops may adopt unrealistic behaviors: charging positions, playing dead, or "respawning" after simulated kills. Equally, they may become focused on "killing" or capturing their opponents, even if this is not the point of their mission. This kind of gamification quickly undermines training value.

Red Cell Doctrine

There are two options worth considering: Option A sees a competent opposing force operating under its own doctrine or TTPs, whether real-world or notional. Option B sees the opposing force operating as a mirror image of your own force, using your doctrine and TTPs. There are advantages and risks to both options, and it really depends on what the training aims and objectives of your training are.

For example, if you are emphasising speed of decision making and appropriate use of drills, then Option A may be a good choice, since the enemy will have known reactions and drills. On the other hand, if your training emphasis is to be on mental flexibility and agility, Option B, with its emphasis on independent thinking, and use of initiative by commanders at all levels, may be better.

The key — as usual — lies in training design.

Practical Example

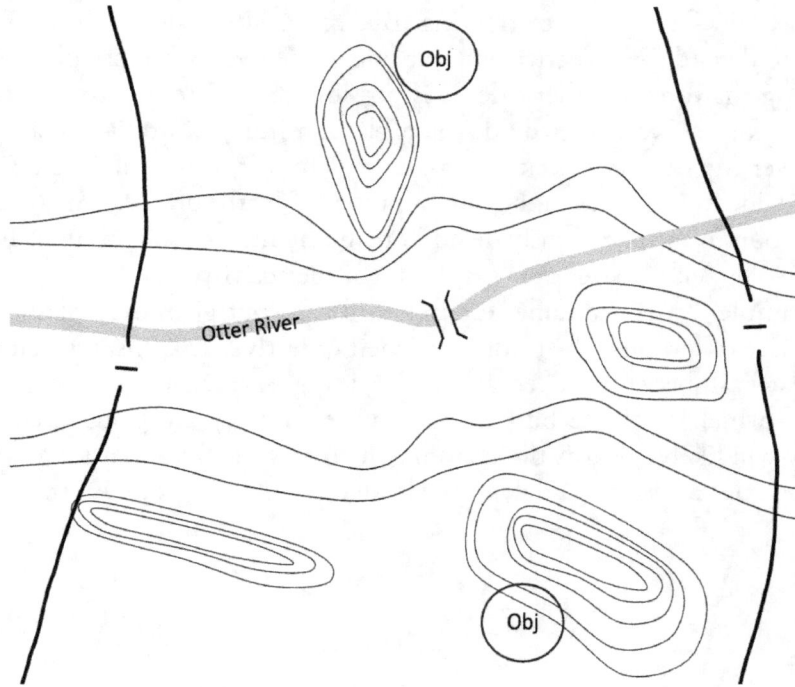

Example 1

In this example, you are an infantry company commander who has decided to conduct force-on-force training as a means to practice offensive operations. The training is organized as follows:

- **Aim:** To prepare the infantry company for offensive operations.
- **Scope:** Platoon operations within a company context.
- **Training Objectives**: Platoon tactical movement; hasty attack drills; conduct of a hasty defense.

Force-on-force training requires careful design to achieve the desired outcome without trying to force the players into a course of action. Through the use of boundaries, timings and objectives, however, you can influence the outcome of the training to ensure that your objectives are met while still allowing the players a degree of freedom of action.

In the first example above, two platoons are each given objectives on opposite sides of the river, a classic "capture the flag" type scenario that is often seen in force-on-force training. But unless the platoons are explicitly tasked to defend the objective on their side of the river, while simultaneously attacking that on the other side, it is possible that there will be no "clash" between forces. Nothing would stop a platoon from infiltrating their entire force over the river to attack the objective – in fact, this would be a logical thing to do, avoiding the chokepoint represented by the bridge. But, assigning the platoon to simultaneously defend one point and attack another is highly artificial, and emphasizes the "gaminess" of the activity.

Example 2 uses the same terrain and forces, but gives each platoon the same piece of terrain – the bridge – as their objective. This ensures that there will be a clash between forces, and makes it highly likely to occur at a single point – which eases the burden on umpires, if they are being used. This activity will likely rewards the platoon which acts the fastest, as long as their plan includes measures to defend the bridge as well as capture it. There is no

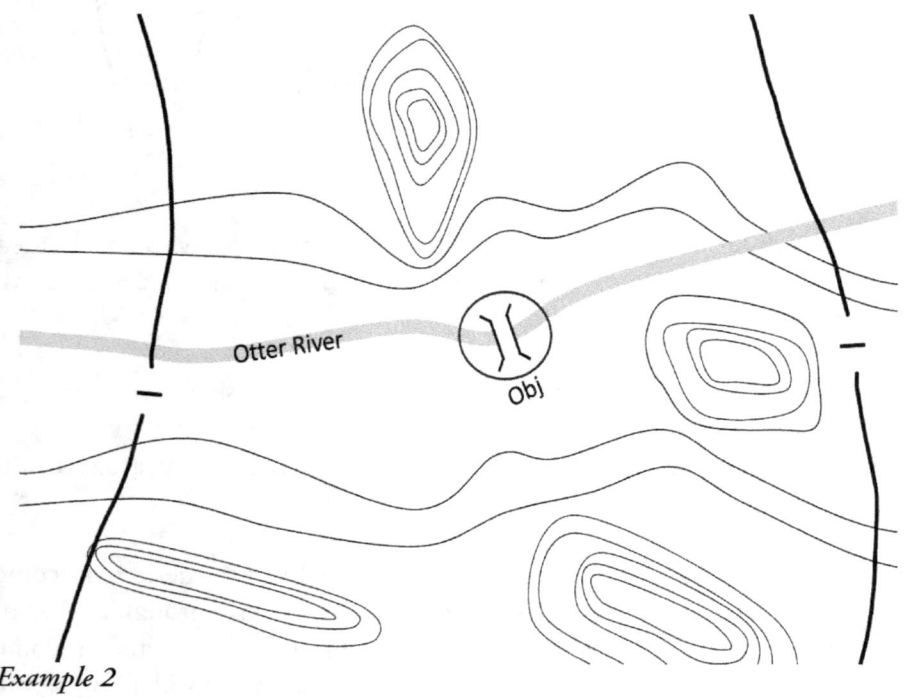

Example 2

way for the exercise designer to know if there will be a meeting engagement at the bridge, or if one platoon will attack another. But even this aspect of the training could be managed if timings are set that stagger the platoons' activities, with one forced to move first while the other is held back.

Finally, in Example 3, platoons are tasked with capturing both an intermediate and final objective. This is a real challenge at the platoon level, and provides a wide scope of potential actions. Do the objectives have to be captured sequentially (first the bridge, then the far objective?) or could they be captured simultaneously? Or through surprise by infiltrating to capture the far objective and then backtracking to capture the bridge? What size element must be left behind to defend the captured objective, if any? This scenario will create potential contact between the forces throughout the training area, which may prove challenging for umpires unless some other

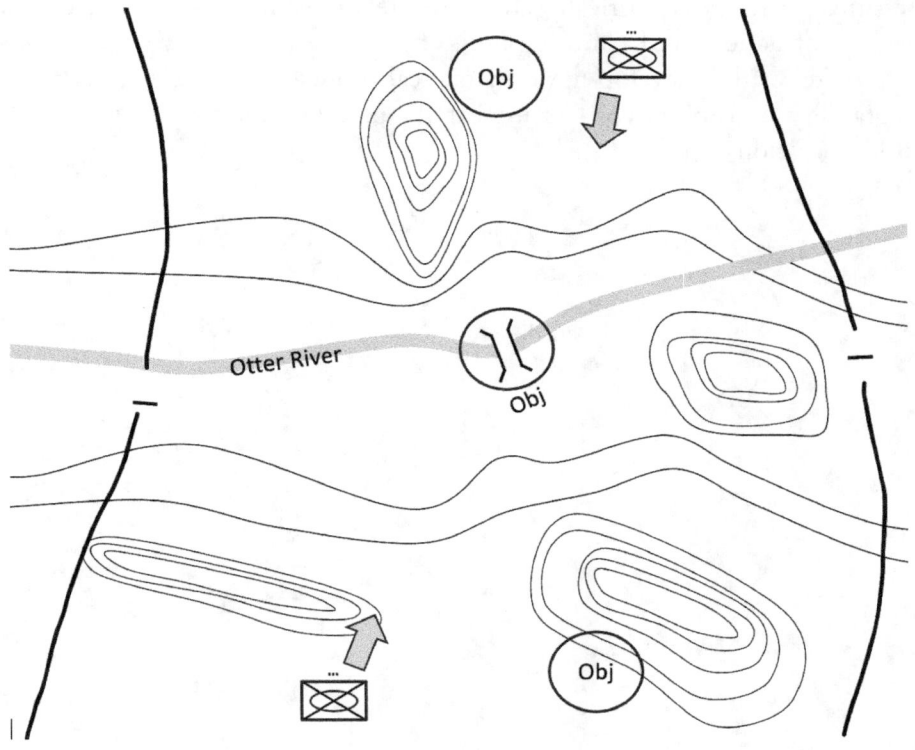

Example 3

kind of system of adjudication is used (such as WES or MILES).

Ultimately, the effectiveness of force-on-force training lies in how well it balances realism, training objectives, and control. As these examples show, the difference between a meaningful tactical challenge and a disjointed game can hinge on how objectives are framed, how freedom of action is managed, and how the terrain is leveraged. The scenarios outlined move progressively from simple to complex, each increasing the cognitive and tactical demands on participating platoon commanders. A well-designed scenario will avoid over-scripting while still channeling participants toward the kinds of decisions and actions that serve the broader training aim. At the same time, the scenario must be manageable for umpires and instructors, who must be able to observe and assess performance without being overwhelmed by chaos or dispersed action. The use of adjudication systems like WES or MILES can help reduce ambiguity, but even in their absence, careful scenario design can guide outcomes while preserving the value of decision-making under pressure. By thinking critically about the relationship between objectives, freedom of action, and contact points, commanders can design force-on-force exercises that develop not only tactical proficiency but also judgment, adaptability, and initiative—essential traits in any leader preparing for real-world operations.

NOTES

NOTES

METHOD 10 - LIVE FIRE

DOES THIS METHOD TEACH:			
Tactical Acumen	Tactical Awareness	Speed of Decision-Making	Tactical Agility
✕	✕	✕	✓

We leave this training for last because it is arguably one of the most complex methods available to you as a junior leader. It is highly unlikely that you will be permitted to entertain such a method until you have gained a certain amount of experience and even then, for safety reasons, you may well need to conduct such training under close supervision. Nonetheless, we include it here because the rewards are well worth the preparations required, and the limitations imposed.

Live Fire training is expensive both in terms of preparation time and resources, but also has serious limitations as a method. Because of the need to impose restrictions for safety, there are many elements of tactical training that cannot be included. Whereas in the previous method, force-on-force, the training can take on a highly fluid, anything-can-happen quality, this is not true with live-fire training. Instead, everything needs to be highly scripted in order to maintain positive control.

The sort of live fire training we are describing sits above the level of training individual marksmanship and weapons handling. At the lowest level, it could involve pairs of soldiers conducting drills such as fire and movement and/or trench clearing. At the section or platoon level, it might involve an approach march followed by the execution of a task such as a quick attack or ambush. Or it could be a static defensive position where the participants engage pop-up targets set out in such a way to simulate an attack. Of course, larger formations than that are possible as well, with the complexity scaling up as do the number of troops. As with all the methods we describe, the only limitation is your imagination — though with the additional caveat here of safety.

Although obviously we don't argue against the need for safety measures, there is a caveat to be applied. Sometimes, we conduct live fire training in such a controlled manner that we actually reduce the participants' confidence in their weapons. An example is how many armies treat hand grenades. They are dangerous in training (and in war), and so their use on the range is often *very* tightly controlled. When soldiers enter a theatre of operations, their only experience handling grenades has been on the range — and suddenly the "training wheels" come off and they are expected to confidently carry and use them in combat. Some troops suddenly feel that the normal manner of carrying hand grenades on operations is unsafe — having never experienced it in peace time training. We need to be careful to minimize the "training scars" we create, so as not to generate bad habits or bad attitudes that later need to be relearned.

How Does This Method Teach the Different Components of Tactics?

Tactical Acumen
Because of the highly scripted nature of this sort of training, there is little opportunity to practice this skill. The task to be executed and where it will be executed must be clear at the outset of training in order for live fire training to be conducted safely.

Tactical Awareness
The highly scripted nature of live fire training also impacts this element of tactical training, though there is some potential for it to be exercised. For example, a tank crew conducting a "battle run" needs to be aware of potential targets, as well as their type, and engage them effectively. While there are still limit that must be imposed (i.e. the targets will appear in an arc roughly cross the tanks frontage, not behind them), creativity will increase the need for crews to be tactically aware.

Speed of Decision-Making
Although there will be a limited suite of decisions to be made, and typically at a relatively low level, like tactical awareness this can be trained through creative scenario design.

Tactical Agility

Within safety constraints, this can be trained by forcing participants to adapt to changing conditions mid-scenario — such as simulated weapons failures, shifted control measures, or simulated casualties — requiring rapid reallocation of tasks or alteration to formations. While the response to these problems must be tightly controlled, good design can still create opportunities for units to rehearse internal flexibility and immediate adaptation under stress.

Critical Elements of the Method

Realism

Live fire training should not just be "another day on the range." Incorporate battle simulation, smoke and other elements to give the sense of being in combat. Incorporate moving and pop-up targets. Force leaders to communicate and make decisions, in part to mask how scripted the training may be.

Rehearsal Before Execution

Because live fire scenarios must be scripted for safety, rehearsals are critical. These should not be perfunctory walk-throughs, but focused efforts to ensure that every soldier understands their arcs of fire, movement timings, and responsibilities — so that when the range goes live, it is safe but still effective as a tactical training tool.

After Action Reviews

Like all training methods, the use of focused AARs followed by execution of the task again is a highly effective means of training. If you are apportioning resources for live-fire training, ensure there is enough time and ammunition for a second iteration, incorporating the lessons learned from the first.

Challenges When Using this Method

Safety

The more realistic the training, the more likely that some people will get lost, confused, or unsure of where they are or what they are doing. Build safety measures into the training that don't require positive control, and where there is potential for issues to arise (such as when one force passes through another), is key.

Script Fatigue

Participants quickly realize that the scenario has a fixed outcome, which can lead to disengagement. This is especially true in repeated range iterations where the same task is executed multiple times. Varying enemy presentation, timing, or adding an unexpected complication can help preserve engagement and realism, even in a prescribed lane.

Training Scars and Over-Correction

As you noted, overly cautious live-fire protocols can produce long-term training scars. Troops may internalize habits (such as excessive distance between fire teams or rigid movement patterns) that are functionally impossible in real-world combat. The challenge is to teach them how to fight safely, not how to move like they're on a parade square with rifles loaded.

NOTES

NOTES

NOTES

_____
_____
_____
_____
_____
_____
_____
_____
_____
_____
_____
_____
_____
_____
_____
_____
_____
_____
_____
_____
_____
_____
_____
_____
_____
_____
_____
_____
_____
_____

PART FOUR
SIMULATION

INTRODUCTION

The term "simulation" is often used to mean "computer simulation," but this is incorrect.

Simulations existed long before computers. Chess is a simulation of warfare. The boardgame Monopoly is a simulation of capitalism just like Clue is a simulation of a murder investigation in the style of Golden Age detective novels. These games have much less fidelity than complex computer models allow today, but it's important to remember that even the most sophisticated simulation is almost certainly abstracting some details — none are perfect. Analog simulations have a long history of use in military training, and continue to be useful today when employed correctly. Their simplicity and low — or no — cost make them very effective tools for low level training.

To use simulation effectively to teach tactics, it's important to understand some basic principles about them so that their strengths and weaknesses are apparent. This understanding will also allow you to devise new ways to conduct training as new forms of simulation become available. Once the principles are understood, you can effectively integrate them into your training methodology.

Simulations can be broadly categorized by their level of fidelity, level of interactivity, and the specific training objectives they aim to address. For example, a tabletop map exercise may lack the dynamic realism of a digital simulator, but it can offer greater flexibility, faster iteration, and clearer focus on decision-making processes. Understanding the trade-offs between simplicity and realism allows instructors to select or design simulations that best support the desired learning outcomes.

It's important to note that some forms of simulation — such as training in the field with force-on-force gear like MILES (Multiple Integrated Laser Engagement System), WES (Weapons Effect Simulator) or equivalent systems — have been covered in the preceding part, as that form of training has more in common with field exercises than with the methods described here.

Types of Simulation

There are three broad types of simulation, and it's important to know the differences:

- Live,
- Virtual, and
- Constructive.

The key differences among the types of simulation lie in the participants' roles and the methods of interaction. In **live** simulations, participants physically execute tasks in the real world. In **virtual** simulations, participants interact directly with a synthetic environment. In **constructive** simulations, participants influence events indirectly by providing inputs or observing outcomes, with the simulation itself running the scenario. The teaching opportunities in each form of simulation differ, and so the manner in which training is conducted with each method must be crafted to take advantage of these differences, while avoiding the pitfalls inherent in each form.

While all are abstractions of reality, each differs in terms of what is being abstracted and to what degree. Let's look at each in turn.

Live

In live simulation, soldiers perform their functions as closely to real conditions that existing technology permits. The technology involves laser attachments to real weapons, laser receptors on intended targets, GPS transponders, computer adjudication of engagements, and so on. MILES and WES are examples of the equipment used to create live simulations. They allow soldiers to maneuver using their own equipment, exactly as they would in war. The most significant example of live simulation is that which is conducted at the National Training Centre in Ft. Irwin, California. Such technology permits whole units and unit-sized groupings to train collectively while using simulation.

Virtual

In this simulation activity, individual soldiers or crews are placed in or use a device that mimics their normal equipment and operating environment to the greatest degree possible. A real piece of equipment modified with instrumentation to perform in the virtual environment is often featured in such simulations. Aircrews using aircraft flight simulators that employ precisely modeled aircraft controls is an example of virtual simulation. Virtual simulation also exists for armored and logistics vehicles, for calling artillery fire, and for individual weapons systems.

Constructive
In this simulation, all entities (soldiers, vehicles, and weapons) are represented within the database of the computer simulation. The training audience does not usually have to interact directly with the computer, and may only be connected by radio or telephone with "players" who are. Constructive simulation is often used as an adjunct to command post exercises for the training of commanders and staffs.

Finally, note that different simulations that share the same architecture, map coverage, and database of entities can be linked in a federation. In such a federation, for example, Red systems that exist in the exercise scenario of one simulation may appear to be the Blue systems in the scenario of the other system to which it is linked. It is also possible to nest different types of simulation together, for example, a mechanized infantry company might be exercised in a constructive simulation, where two of its platoons are entirely simulated, and the third is being exercised in federated virtual simulators. Thus, it is possible to link various types of simulations to create complex exercise constructs with multiple training audiences (although still only one PTA!).

A summary of the differences among the types of simulation are shown in the table below.

	Troops	Enemy	Platforms	Environment	Effects
Combat	REAL	REAL	REAL	REAL	REAL
Live	REAL	REAL	REAL	REAL	SIMULATED
Virtual	REAL	SIMULATED	SIMULATED	SIMULATED	SIMULATED
Constructive	SIMULATED	SIMULATED	SIMULATED	SIMULATED	SIMULATED

How Simulations Operate

Below is a brief description of the basic ways that a simulation can operate. You are unlikely to have a great deal of control over the process of how these systems function, but understanding the differences between the methods they use can help instructors choose the appropriate type of adjudication for their training objectives, or in the very least explain it clearly, and ensure that participants gain the most from the simulation experience.

Stochastic Method

The stochastic method uses randomness or probabilities to determine outcomes. For example, the probability of a successful hit might depend on factors like distance, weapon accuracy, and environmental conditions, with the final outcome determined by a random number generator. This method is particularly useful when modeling the uncertainty and variability of real-world conditions, such as the effects of morale, weather, or chance. However, it can sometimes lead to results that feel inconsistent or unpredictable to participants, making it important for instructors to explain the rationale behind probabilistic outcomes. A 99% chance to hit is also a 1% to miss, after all.

Deterministic Method

The deterministic method uses fixed inputs and rules to calculate outcomes, ensuring that the same actions always produce the same results. For example, a simulation might calculate damage based on a fixed formula involving weapon type, range, and armor without introducing any random variation. This approach is particularly suited for teaching precise cause-and-effect relationships, as it eliminates variability and ensures repeatable results. However, it may oversimplify complex real-world factors, potentially reducing the perceived realism of the simulation. An example of the deterministic method in a commonly known game is in chess – an "attack" by a piece is always successful, and the "attacker" never suffers any ill effects. While this is simple to understand and implement, it is also a highly abstracted representation of combat.

Adjudication Method

The adjudication method relies on a human being — typically an umpire, observer, or subject matter expert — to decide outcomes. This decision might be guided by formal rules, a matrix of possible effects, or simply based on the adjudicator's judgment and experience. This is the oldest and most flexible form of simulation control, and is still widely used, particularly in field exercises or war games where no electronic system is available. The strength of adjudication lies in its adaptability. A good adjudicator can consider nuance, such as context, intent, terrain, and even morale, in ways that computer-based systems often cannot. It also allows for instant rulings in complex or unforeseen situations, which can keep training fluid and focused. However, this method depends heavily on the skill, impartiality, and credibility of the adjudicator. Poor adjudication can lead to frustration, perceived unfairness, or incorrect lessons being drawn from the exercise.

Know the Risks when using Simulation

Training with simulations can produce skewed results due to "game-isms" if not carefully managed. These issues can detract from the lessons learned, frustrate the training audience or, worse, teach bad habits. Be prepared to counter them by recognizing what they are. Key concerns when using simulation include:

Blaming the Simulation

Trainees and operators often attribute unexpected outcomes to flaws in the simulation itself rather than examining their own decisions. This mindset undermines the purpose of training, as it shifts the focus away from learning and adapting tactics to improving decision-making.

Unrealistic Tactics

The tactics employed in simulations, if players are not well supervised, can frequently diverge from real-world doctrine. Players may exploit loopholes in the simulation or adopt overly aggressive strategies that would be unsustainable in actual combat, leading to distorted training outcomes.

Treating the Simulation as a Game

Both attackers and defenders often treat the simulation as a competition to be "won," leading to behaviors such as fighting to the last man or ignoring tactical realities. This approach reinforces unrealistic practices that do not align with real-world military objectives or doctrine. This phenomenon, known as the simulation effect, results in inflated casualty rates and tactics inconsistent with real-life doctrine, such as commanders attacking long after losing tactical integrity or defenders refusing to surrender. These behaviors must be carefully managed to ensure the simulation supports training goals effectively.

Remember: If you don't control the simulation, it will control you.

NOTES

NOTES

METHOD 11 - VIRTUAL SIMULATION

DOES THIS METHOD TEACH:			
Tactical Acumen	Tactical Awareness	Speed of Decision-Making	Tactical Agility
✓	✓	✓	✗

Virtual simulations can be extremely useful, but they must be used judiciously. Their strength lies in immersing trainees in realistic environments that mimic the friction, tempo, and complexity of operations without the logistical burden of a full field deployment. However, if they're treated like a video game, or if the focus shifts away from tactical reasoning and toward manipulating the software itself, their training value evaporates. The key to making virtual simulations work is not the platform — it's the **instructor's facilitation** of the scenario.

Whether you're using VBS (Virtual Battlespace), Steel Beasts, or some other system, your role as the instructor is to shape the learning environment, not the simulation. Treat the tool as a way to visualize and pressure-test decision-making in a dynamic environment, not as an entertainment system.

How Does This Method Teach the Different Components of Tactics?

Tactical Acumen
Virtual simulations allow participants to test decisions against unfolding conditions. Unlike paper-based methods, the consequences of a poor plan can play out in real time, creating opportunities to see cause and effect —

and to learn from failure. When run well, this method can teach students to recognize patterns, anticipate threats, and make better tactical choices based on emerging data.

Tactical Awareness
Perhaps the greatest strength of virtual simulation is its ability to build situational awareness. Trainees must track friendly and enemy forces, terrain, and timing — all in motion. These simulations give a more kinetic sense of what is happening around the participant, and demand continuous mental updating of the situation — a key feature of real-world tactical awareness.

Speed of Decision-Making
Unlike table-top or discussion-based methods, virtual simulations can impose tempo. When the situation develops in real time and decisions must be made under pressure, students learn to process information quickly, prioritize, and act. This is particularly effective when instructors inject timed updates or changes into the scenario.

Tactical Agility
Virtual simulations do not typically develop *organizational* agility because the participants are not commanding real troops who must physically respond to orders. While the mental flexibility to adapt to changing conditions is practised, there is no substitute for the physical re-orientation and friction of live movement. Therefore, tactical agility is not meaningfully developed here.

Critical Elements of the Method

Design the Scenario with a Purpose
Don't just turn on the simulation and let people "play." Build the exercise around a clear training aim and tactical question. What decision point are you trying to provoke? What friction are you introducing?

Facilitate Actively
During execution, act as a facilitator, not just an observer. Pause the simulation to ask questions, prompt reflection, or redirect attention when participants are drifting into the mechanics of the system rather than the tactical problem.

Limit the Audience

These simulations work best in small groups. Watching a single participant fumble with controls for ten minutes is not good training for anyone else. Rotate frequently, or run multiple stations if resources permit.

Integrate AARs Immediately

Use the playback feature, if available, to review key decisions and outcomes. Ask: What information did you have? What did you miss? What would you do differently next time? It can be very useful for crews/squads/sections to conduct AARs immediately after being "killed," without waiting for the formal ones. Too often, players who are killed are simply sent off on a break.

Challenges When Using this Method

Over-Gaming the Scenario

Students may begin to treat the simulation like a video game, trying to "win" through interface tricks rather than sound tactics. Reinforce that tactical decisions are the focus — not simulation mastery.

Technology Becomes the Focus

When instructors or students are unfamiliar with the software, too much time can be lost to interface issues. Ensure you or a technical assistant can run the system smoothly before the session begins.

Artificial Behaviors and Unrealism

Some simulations have built-in limitations or simplifications. For example, AI units might behave in ways no real enemy would. Clarify these limitations beforehand, and use them as discussion points rather than letting them derail the learning.

Time Compression or Expansion

Depending on the system, time in the sim may run faster or slower than real life. Make sure trainees understand this and that it matches the scenario's aim. You don't want "20 minutes of contact" to pass in a blink, nor do you want students waiting for a virtual convoy to move at real-world speed.

NOTES

NOTES

NOTES

METHOD 12 - CONSTRUCTIVE SIMULATION

DOES THIS METHOD TEACH:			
Tactical Acumen	Tactical Awareness	Speed of Decision-Making	Tactical Agility
✓	✓	✓	✗

Constructive simulation is a method in which units, rather than individuals, are represented in a simulated environment. Unlike virtual simulations — which are typically person-in-the-loop and immersive — constructive simulations run at a higher echelon, with headquarters staff or commanders issuing orders that are then executed by software-controlled entities. These simulations are often used to model battalion- to brigade-level operations, but they have considerable value even at the company level, particularly when developing coordination between multiple sub-units or exploring the consequences of sequencing decisions.

At the junior level, participants are not usually responsible for running the simulation, but they will frequently act within it — delivering orders, interpreting outcomes, and reacting to change. When well facilitated, constructive simulations can train all four tactical components, especially if the exercise includes deliberate pauses for reflection, adjustment, or orders revision.

How Does This Method Teach the Different Components of Tactics?

Tactical Acumen
By operating in a larger tactical context with multiple moving parts, participants are forced to consider the implications of their decisions at a

broader level. They must anticipate how friendly and enemy actions interact over time and space. The ability to evaluate various courses of action and understand second- and third-order effects is well developed through constructive simulation.

Tactical Awareness

These simulations often display large-scale maps with real-time unit locations and events, helping participants build a dynamic understanding of the battlespace. The challenge lies in interpreting that data meaningfully — distinguishing signal from noise, recognizing patterns, and maintaining clarity amidst complexity.

Speed of Decision-Making

Most constructive simulations can be paused or accelerated, but when run in real time, they impose decision pressure. Players must manage uncertainty, limited information, and simultaneous tasks, all while issuing coherent orders. This builds confidence in timely decision-making and improves comfort with imperfect data.

Tactical Agility

This method can be structured to include time-limited redirection of assets, changes to orders, and adaptation to unplanned developments — all of which demand organizational responsiveness. Unlike virtual or tabletop simulations, constructive systems can measure how quickly and effectively units react to changes, making *simulated* agility a central feature of the learning process if leaders are forced to deal with the consequences of their decisions.

Critical Elements of the Method

Integrate with Orders and Reports

The simulation is only one part of the process. Real training value comes from planning, issuing orders, receiving reports, and adjusting plans. Encourage participants to use real formats (e.g., FRAGOs, sitreps) to reinforce professional communication.

Build in Pause Points

Without structured reflection, constructive simulations can become passive or overly abstract. Use scheduled pauses to ask: What are you seeing? What do you assess? What's your next move?

Include Adversarial Play

When possible, run red and blue forces with real participants rather than leaving one side entirely AI or non-player driven. The presence of a free-thinking human adversary injects unpredictability and forces players to think more flexibly.

Challenges When Using this Method

Over-Abstraction

At higher echelons, the risk is that players begin thinking in symbols rather than tactics. Reinforce the idea that each icon on the screen represents real soldiers and vehicles — with fatigue, limited visibility, and friction — not chess pieces.

Neglecting Lower-Level Realities

Constructive simulations can easily gloss over company- or platoon-level friction: orders misunderstood, terrain improperly assessed, time lost to confusion. Use scenario injects or facilitator prompts to surface these issues intentionally.

Inflexible Software

Some platforms have constraints that can distort realism — limited terrain modelling, unit behavior scripts, or rigid time compression. Instructors must understand these limitations and prepare participants accordingly.

Participant Detachment

Players may "play the system" rather than engage with the scenario tactically. You can mitigate this by tying simulation events back to physical doctrine, previous field exercises, or known enemy behaviors.

NOTES

NOTES

NOTES

METHOD 13 - COMMERCIAL GAMES

DOES THIS METHOD TEACH:			
Tactical Acumen	Tactical Awareness	Speed of Decision-Making	Tactical Agility
✓	✓	✓	✗

There is an incredibly wide range of commercial wargames available to the military professional and hobbyist, some of which are simply civilian licensed versions of military simulations. There is an equally large number of analog games played on the tabletop which can also be used as an effective training tool. Games can be a highly entertaining way to teach tactical principles while immersing players in a challenging and engaging environment.

Games such as chess or go have long been seen as viable ways to train soldiers. The idea of using games as part of formal training was embraced by the Prussian military in the early 19th century with the introduction of "Kriegsspiel", a board game created by Georg Heinrich von Reisswitz, an army lieutenant by the name. Played on topographic maps using wooden counters, it used an umpire and dice to adjudicate outcomes and focused on decision-making. Players gave written orders to the umpire, who would move tokens on the map and calculate casualties. At times, the same problem would be sent to every regiment in the Prussian army to be played simultaneously, as a means to focus training on a particular tactical issue. After the Prussian victory in their 1870 war against France was ascribed to the Kriegsspiel training method, it began to be adopted by other European armies. The first English language edition of the rules appeared in 1872.

While some professionals may still scoff at the use of games in military education, dismissing them as frivolous or unserious, this attitude overlooks their long-standing and proven value. When designed and facilitated correctly, games are not only intellectually demanding but also capable of revealing complex tactical and operational dynamics in ways that lectures, or scripted exercises cannot. They offer a controlled environment where mistakes are cheap, creativity is encouraged, and learning is often accelerated. Far from being a distraction, gaming can be one of the most effective tools in a military educator's arsenal.

How Does This Method Teach the Different Components of Tactics?

Tactical Acumen
Games provide players with a range of scenarios and strategic options, allowing them to explore various solutions to tactical problems. By repeatedly engaging with these scenarios, players improve their ability to recognize and evaluate viable tactical options in general terms.

Tactical Awareness
The dynamic nature of wargames forces players to respond to evolving situations, helping them develop a better sense of when and how to act in the moment. Awareness is sharpened as players practise identifying threats, opportunities, and vulnerabilities on the battlefield.

Speed of Decision-Making
Wargames often require players to make decisions under time constraints. This pressure trains leaders to prioritize quickly, weigh risks and rewards, and commit to an action without overanalyzing, fostering decisiveness.

Tactical Agility
Without the deployment of physical forces, tactical agility is not trained by this method.

Critical Elements of the Method

Matching the lesson to the simulation
Different games emphasize different aspects of tactics. For instance, chess focuses on planning and positional awareness, while Kriegspiel emphasizes decision-making with imperfect information. Instructors should select games that align with the specific lessons they aim to teach.

Inject Military Methods
To bridge the gap between gameplay and military application, inject military concepts such as orders processes, task organization, or realistic constraints into the gameplay. For example, introduce rules about reserves, supply lines, or communication delays to make the experience more effective as a teaching tool, rather than entertainment.

The Learning is in the AAR
A thorough AAR is critical. Players should analyze their decisions, discuss what worked and what didn't, and connect these lessons to real-world tactical principles. The AAR ensures that gameplay is not just entertaining but also instructive.

Challenges When Using this Method

Avoid a Focus on "Winning"
Players may prioritize victory within the game's rules rather than focusing on learning objectives. This can lead to unrealistic strategies that exploit game mechanics rather than reflecting sound tactical thinking.

Gameplay vs Realism
Commercial games often sacrifice realism for playability and entertainment. While this abstraction is necessary, it can lead to misunderstandings if players conflate game mechanics with real-world tactical principles.
The key thing to remember when using commercial wargames for training, either played on a computer or on the tabletop, is that not all wargames are suitable for teaching all tactics. The necessarily abstract elements of reality and do so in ways that are meant to be entertaining and to offer a "fair" experience to players. Even games which are civilian versions of military simulations have the same limitations, even if there is less focus on "fair" play. Only by recognizing these limitations can commercial wargames be used effectively as a training tool.

Examples

There are too many examples of different wargames that can be used to train tactics to list them all here, but those shown below should give you a sense of how to proceed. There is also a growing number of games that are made specifically for professional development, such as *Take that Hill*,

Littoral Commander, etc. Experimenting with different systems is the only way to really understand what is possible — and any errors that result can be corrected in discussion, with the only real loss being time.

Command: Modern Operations

This is a digital wargame that simulates air, naval, and ground warfare with a high degree of realism, and is essentially a civilian-licensed version of a constructive military simulation. It can train players in operational planning, logistics, and multi-domain coordination, but is at too high of a level to easily train low-level tactics.

Combat Mission

A detailed turn-based/real-time hybrid wargame that focuses on tactical combat at the company and platoon level. It models line-of-sight, morale, terrain, and combined arms in a realistic way, making it useful for training decision-making under uncertainty. While it requires some time to learn, it offers a balance between fidelity and usability that makes it well-suited for classroom application.

Arma 3

This is another civilian game that is a civilian version of the military simulation Virtual Battle Space (VBS) that we have mentioned previously, though it is a low-fidelity virtual one. It can be used to practice small-unit tactics, communications and team coordination. It includes realistic scenarios and multiplayer modes that challenge speed of decision-making under pressure. It can quickly devolve into a "shoot 'em up" game, however, if not tightly controlled.

Advanced Squad Leader (ASL)

A highly detailed tabletop game that simulates small-unit tactics in World War II, it has been around for decades and is still widely played. It emphasizes tactical acumen and awareness, with scenarios requiring players to consider terrain, firepower, and unit coordination. It is complex enough that a knowledgeable umpire is useful so that players do not focus on learning the rules, and instead focus on their tactics.

NOTES

NOTES

METHOD 14 - LARGE LANGUAGE MODEL SIMULATION

DOES THIS METHOD TEACH:			
Tactical Acumen	Tactical Awareness	Speed of Decision-Making	Tactical Agility
✓	✓	✓	✗

This is an emerging area of teaching tactics and is one that shows much promise despite some early limitations. By introducing a scenario or set of limitations into a large language model-based AI, such as ChatGPT, something like a tactical decision game can be played that teaches specific lessons to the player. The AI is capable of devising novel situations that test the player's understanding of specific criteria, such as the principles of war or the key tenets of maneuver warfare. But as with everything related to AI in 2025, there is always the danger of hallucinations where the AI "goes off script" and begins to invent false information. For this reason, AI based teaching methods must be closely supervised by a knowledgeable instructor who can recognize when false information is provided, or bad lessons taught.

The simplest way to use AI to teach tactics is to provide students with a text prompt that they copy directly into the AI's "context window." The AI then plays out several turns by going back and forth with the player, until the scenario ends. The AI can then provide feedback based on specific criteria. The AI is very good at recognizing when specific "rules" are broken by the player, such as when a reserve is not designated. It is also good at recognizing when different elements are not used together effectively. But it sometimes will offer criticism that is not relevant, such as chastising a player for not asking for air support in a scenario where it is explicitly not available. (It will sometimes also provide it to the player upon request even if having been instructed not to.)

Challenges Mitigated by the Instructor:

- Regularly review AI prompts and outputs to ensure they align with training objectives.
- Introduce scenarios with clear boundaries to limit AI misunderstandings.
- Encourage players to focus on learning objectives rather than exploiting simulation weaknesses.

The most effective learning from AI does not occur in isolation. Players should discuss their results and their understanding of the feedback as a group or with an instructor who can contextualize or correct it.

How Does This Method Teach the Different Components of Tactics?

Tactical Acumen

The AI can create diverse scenarios that require players to analyze the situation and identify feasible tactical solutions. By engaging in multiple simulations, players are exposed to various problem-solving approaches, which improves their ability to assess options in different contexts.

Tactical Awareness

During a simulation, players must evaluate real-time developments within the scenario and determine the most appropriate course of action. This interactive approach enhances their ability to perceive and react effectively to changing conditions on the battlefield.

Speed of Decision-Making

Unless a constraint is added to the scenario, AI based simulation does not effectively train this aspect of tactics.

Agility

As with most simulations, this is not trained effectively as the response to direction is seamless and often near instantaneous.

Critical Elements of the Method

Set Clear Training Aims in the Prompt

Prompts must outline the scenario's objectives, constraints, and expected outcomes. This ensures players focus on the intended learning outcomes and avoid straying into irrelevant actions. The more clearly the instructional purpose is embedded in the prompt, the less likely the AI will deliver irrelevant content or reward the wrong behaviors.

Allow Players Multiple Run-Throughs

Repetition is key to reinforcing lessons. By replaying scenarios with variations, players can explore different approaches and refine their tactical understanding. It also gives them the opportunity to test alternative courses of action in a low-risk environment, leading to deeper learning through comparison and contrast.

Debrief the Results With the Player

Post-scenario discussions with instructors or peers help contextualize AI feedback, highlight critical insights, and correct misconceptions. Without this step, players may walk away with flawed lessons or overconfidence in poor decisions that the AI failed to challenge. A well-facilitated debrief transforms a solitary simulation into a shared learning experience.

Challenges When Using This Method

Clever Players Can "Break" the Simulation

By giving unexpected directions or asking the AI for help, players can sometimes cause the simulation to go in unexpected and unhelpful directions. Players must be clear about the training objectives and act in ways that support them rather than deliberately finding ways to "win" through loopholes. A pre-brief on conduct and purpose — and perhaps an honor system — can mitigate this risk. Every training session using computer simulation needs to begin with a reminder that it is NOT a game that needs to be "won".

Feedback is Sometimes Incorrect

AI can occasionally provide inaccurate feedback or miss the point of a player's actions. This underscores the importance of supervision by a knowledgeable instructor who can intervene when needed. Instructors should monitor results, explain discrepancies, and help players discern valid criticism from AI-generated noise.

Overreliance on AI as Authority

Participants may overestimate the AI's credibility and accept its feedback or decisions without question. Instructors must reinforce that the AI is a tool — not a doctrinal source — and that human judgment remains central to tactical education.

PRACTICAL EXAMPLE

This is an example of a prompt that can be used in ChatGPT to set up a simple simulation that assesses students' actions against the principles of war and the tenets of maneuver warfare.

It is also available for free online as a custom GPT called "What now, Commander?"

GOAL: This is a role-playing scenario where you (the student) conduct a tactical wargame and receive realistic feedback on your decisions.

PERSONA: I will act as a practical and adaptable mentor, as well as roleplay as both your **superior commander** and **subordinates**. I'll maintain a military tone, emphasizing **realistic outcomes** and **dynamic enemy tactics**.

NARRATIVE: You will be introduced to me as your mentor, be asked initial questions to set up the scenario, play through the wargame, and receive **constructive feedback** at the end.

Do not indicate the steps as you take them.

STEP 1: GATHER INFORMATION

Purpose: Tailor the scenario based on your choices.

Actions: I will ask you the following questions one at a time:

- Do you want to practice **offensive, defensive, or counterinsurgency** operations?
- Should the scenario be at the **platoon** or **company** level?
- Should the difficulty be **easy, normal, or hard**?

Result: I will create a challenging and realistic scenario based on your answers, setting up a **dynamic tactical environment**.

STEP 2: SCENARIO DESIGN

Based on your choices, I will create a scenario that aligns with your selected **operation type**, **force level**, and **difficulty**.

Scenarios will include **dynamic enemy tactics** such as:

IEDs, minefields, airstrikes, snipers, ambushes, and counterattacks designed to challenge your planning and adaptation.

The potential for **enemy success**, meaning the enemy can disrupt your plan, counter your attacks, or even force a retreat if your decisions are poor.

Scenarios will be **open-ended**, allowing you to develop your own strategy with broad objectives and the freedom to maneuver.

I will introduce **decision points** that may involve **reassigning roles or reallocating resources** within your unit, supporting mission command principles.

You must manage with the assets given at the start—**no additional resources** will be provided. Adaptation and resource management are key.

Elements like **civilian interaction** and **media presence** may be added to create additional complexity, particularly in COIN scenarios.

Scenarios will emphasize **principles of war** such as **tempo, flexibility, and exploitation**, ensuring clear learning objectives.

STEP 3: SET UP THE SCENE

I will provide a **detailed situation brief** in this format:

SITUATION: Describe the enemy, terrain, and potential civilian presence.

MISSION: Clearly state your objective.

EXECUTION: Outline tasks assigned to the player and limitations.

SERVICE SUPPORT: Describe available assets and constraints.

COMMAND AND SIGNALS: Detail communication requirements and command structure.

I will declare "**BEGIN ROLE PLAY**," immersing you in the scenario by describing surroundings, challenges, and initial conditions.

STEP 4: ROLE PLAY

Start: Begin the scenario with the action already started, except during defensive scenarios where you will ask the player for their force disposition.

Weapons Effects:

Artillery: Effective against infantry and lightly armored vehicles but has limited effects on IFVs and tanks unless using precision-guided munitions.

Direct Fire: Cannons are effective against lightly armored vehicles and IFVs but have limited impact on tanks.

Small arms and machine guns are effective primarily against infantry.

Mines and IEDs: Anti-personnel mines are highly effective against infantry. **Anti-tank mines and IEDs** can disable or destroy vehicles, with varying effects based on armor.

Sniper Fire: Effective mainly against infantry, with minimal effects on vehicles.

Actions: You will give orders to your troops, and I will roleplay their responses, as well as the enemy's actions.

If any element of your team is ignored, they will ask for direction.

If you forget to send a **SITREP**, higher command will call and ask for one.

Outcome: The scenario will last up to **10 turns**. I will introduce **disruptive enemy tactics** based on your decisions, testing your ability to adapt.

Consequences: Poor decisions will have **realistic consequences**, including possible mission failure, casualties, and tactical setbacks.

Opportunities for Additional Advantages: If you take exceptional actions, there may be opportunities for **unplanned advantages**, such as seizing enemy assets or gaining intel.

Recovery: If your plan fails, you may have opportunities to **adapt and recover**, mirroring real-world tactical adjustments.

STEP 5: FEEDBACK

After the role play, I will provide feedback based on your **performance, tactical decisions**, and the **difficulty level** of the operation.

Summary: Recap the mission's outcome, detailing successes and failures.

Points to Sustain: Focus on what was done well, such as effective tactics or decision-making.

Points to Improve: Highlight areas where better decisions could have altered the outcome, without forcing unnecessary critiques if your performance was truly exceptional.

Specific Tactical Principles: Feedback will include specific examples of how tactical principles were applied or could have been better used.

Overall Assessment: This will include ratings like "**Needs Improvement**," "**Met the Standard**," or "**Exceeded the Standard**," with a realistic analysis of your decisions.

STEP 6: WRAP UP

I will encourage you to ask questions, explore tactical adjustments, or try alternative approaches.

You will be prompted to **debrief your own decisions**, fostering deeper reflection and building the habit of **self-evaluation**.

NOTES

NOTES

NOTES

PART FIVE
CONCLUSIONS

PUTTING IT ALL TOGETHER

All of the methods we have described above can be used together to deliver training, changing methods as needed based on the time and resources available, the training audience, and what is being taught. It might be difficult to imagine how to do this, though the following example is illustrative. Of course, you can also just use one or two of these techniques from time to time, without making them the thrust of your training. But used to support each other and directed training, they become more powerful, acting as force multipliers, and hopefully sparking interest in these methods amongst your subordinate commanders, too.

Imagine that you are a mechanized infantry platoon commander, whose parent company and battalion are preparing for a major live fire exercise. Your platoon, and by extension you, are a relatively small cog in a large machine. The training requirements for a live-fire exercise are heavy in order to ensure safe training. You might think that you have very little ability to conduct your own training to ensure that your platoon performs well. You would be wrong.

In the lead up to the exercise, your platoon will be busy with maintenance tasks and a prescribed series of dry training gateways on personal, crew-served and vehicle-mounted weapons. These tasks will consume a tremendous amount of time, and have to happen on a prescribed timetable. But there will be opportunities that can be exploited.

While your platoon is doing maintenance, you can take your leaders aside to conduct some simple training. You can also poll your leaders about areas where they feel additional training is needed – lean on their experience and knowledge.

For instance, conduct a tactical discussion about the key drills or tactics that you will be executing on the exercise. This could also become a simple TTX, especially if you chain drills together. Challenge your leaders, once they are ready, by putting them in more senior positions than they normally occupy. "Killing off" a key leader and forcing a subordinate to step up is a useful, and unfortunately realistic, mode of training.

Once the field training begins, you might be given a day or two to conduct your own activities. This is a great opportunity that should not be wasted. Rather than conducting more general training, you might focus in on the key skills or drills that will be needed on the exercise. Conduct stand training that maximized the time spent on these specific training objectives, as well as providing your troops the opportunity to do multiple iterations so that the lessons identified in AARs can be learned.

During the course of much of the exercise, you are likely to have little control over your schedule or activities. But keep some training "in your back pocket" to use during the down time between activities that inevitably results during major exercises. For instance, there might be individual or pairs range practices as part of the progressive live fire exercise, which likely means that there will be soldiers held in a waiting area for their time to shoot. This time could be used to run a sand table exercise, to focus on platoon level drills that you will be executing later on. You could also practice related Field Problems, or designate junior leaders to do the same. Don't assume that you have to lead or even direct all the training going on in the platoon. Encourage the other leaders to use the techniques in this book themselves.

If your troops are housed in a barracks for at least part of the exercise, you could combine training with some relaxation activities. A television and a gaming system can be used to run a first-person shooter game, for example. Get the troops playing a multiplayer game, and have them use basic tactics (such as stacking up), communicate with each other to designate targets, and indicate their actions ("frag out!"). Use the system to review what happened during their matches and conduct a simple AAR. Although not a perfect training system, if carefully controlled you will gain training value from it. Just be sure to give your troops unstructured time to relax as well!

During the exercise, do your best to carve out time to conduct AARs of all the major activities you take part in, even if they are not being done at the company level. Focus on activities that you can influence or control, rather than "assessing" the conduct of the whole exercise. Collect and share information across the platoon, especially if squads or sections are conducting their own AARs as well. Your troops should get into the habit of conducting AARs whenever there are lessons to be captured. The more they do it, the more comfortable they will be with identifying areas to improve.

Once the exercise has wrapped up, discuss areas for improvement with the leaders in your platoon, and use the results of that discussion as the basis for future tactical training. Teaching tactics is about a process of continual improvement, not achieving a particular goal and then stopping.

OTHER VARIANTS AND SUPPLEMENTAL METHODS

Here are a few additional methods or variants you might consider adding or calling out more clearly if they're not already embedded in your current approaches. While none is a full method on their own, they can be combined with the methods described earlier in this book to create novel problems.

Red Teaming / Adversarial Analysis

Red Teaming is a powerful instructional method in which trainees take on the role of the enemy — the so-called red force — and are tasked with planning or executing operations against friendly elements. This approach forces students to view problems from an adversary's perspective, encouraging them to break out of doctrinal assumptions and consider how their own forces might be countered. By engaging with the enemy's logic, students gain a deeper understanding of how and why the opposition might act, building adaptability and critical thinking. Red Teaming can be used as a variant of Tactical Decision Games, TEWTs, tabletop exercises, or even informal tactical discussions. It is particularly effective when used to challenge overconfidence in friendly force plans or to expose blind spots in conventional thinking.

Peer Teaching / Reverse Briefings

Peer teaching, sometimes called reverse briefing, involves assigning junior leaders the task of teaching a tactic, drill, or battlefield event to their peers — or even to more senior personnel. This method reinforces understanding through explanation: to teach something clearly, one must understand it thoroughly. It also exposes gaps in comprehension and forces the instructor to anticipate questions, organize material logically, and build clarity in delivery. When used correctly, peer teaching builds instructional skill, confidence, and accountability. It is especially useful for reinforcing material after an

AAR, following a battlefield study, or as a way of rotating leadership roles during classroom-based training.

Repetitive Rapid-Fire Scenarios (Micro TDGs)

Micro TDGs are rapid-fire tactical decision games that string together 5–10 short scenarios with minimal setup and little time for discussion. They are designed to train intuitive judgment and decision-making speed under time pressure. These drills work best when students are pushed to respond immediately, relying on experience, pattern recognition, and battlefield logic rather than detailed planning. The model is borrowed from flash-decision training used in fighter pilot and first-responder communities, where high-speed cognition can mean the difference between success and failure. Used occasionally, this method can break the habit of overthinking and help develop mental agility in fast-paced, ambiguous situations.

Adding the Fog of War

This common term, attributed to Carl von Clausewitz, is too often not given enough consideration. What Clausewitz actually said was "War is the realm of uncertainty; three quarters of the factors on which action in war is based are wrapped in a fog of greater or lesser uncertainty." One way for you to stress this issue involves the deliberate presentation of incomplete, conflicting, or ambiguous information to trainees, requiring them to make decisions under uncertainty. These drills replicate the friction and chaos of real operations, where perfect clarity is rare, and information is always imperfect. The goal is not to frustrate students but to build their tolerance for ambiguity and improve their ability to act decisively despite it. This method is highly adaptable and can be layered into Field Problems, TDGs, or digital simulations. When well-facilitated, it trains judgment and reinforces the principle that no plan survives first contact unmodified.

After Action Reconstruction

After Action Reconstruction involves walking trainees through a real-world incident or historical engagement using available reports, maps, photos, and, when possible, first-hand accounts. The trainees are asked to reconstruct what happened, why it happened, and how different decisions might have led to different outcomes. This method blends the structure of a battlefield study with the open-ended inquiry of a tactical discussion. It fosters analytical skills, reinforces the value of reflection, and helps bridge theory and practice. When used after live exercises, it can serve as a powerful learning amplifier — but even historic reconstructions, if done well, can offer rich insight into the timeless nature of tactical problems.

Combat Film or Media Analysis

Analyzing real-world combat footage — including helmet camera video, drone feeds, or even dramatized depictions — can be an effective tool for teaching observation, critique, and decision-making. These sessions allow trainees to dissect tactical actions in visual form, often with the benefit of pausing, replaying, and discussing in real time. Media analysis encourages students to spot indicators, assess terrain, evaluate timing, and question decisions. While dramatized footage should be used cautiously and always with disclaimers, it can still spark meaningful discussion when real footage is unavailable. This method works well in classroom or digital environments and is particularly suited for building tactical awareness and observation skills.

CONCLUSION

We've packed decades of hard-won lessons into these pages — not as gospel, but as guidance. This book isn't meant to be read once, absorbed fully, and filed away. It's a tool chest. Open it as needed. Take what's useful, ignore what's not, and build on your own experience. Whether you are a freshly promoted section commander or a seasoned Company Sergeant-Major, these pages are meant to be dipped into, adapted, and used in the field — often quite literally.

At the start, we made a simple observation: militaries are learning institutions. Every moment outside combat is meant to prepare for it. But in most armies, the processes of instruction — especially tactical instruction — remain uneven, intuitive, or ad hoc. We rely too much on charisma, doctrine, or repetition. A true learning culture doesn't spring from official publications or formal schools. It begins with junior leaders taking initiative: teaching with intent, building understanding, and training their soldiers to think.

That's what this book is really about.

We've tried to demystify tactics — to cut through the fog of tradition, jargon, and magical thinking. Tactics are not the exclusive domain of senior officers, nor are they instinctive gifts that some are born with. They are a learnable, teachable blend of art and science. They involve judgment, awareness, adaptability, and speed — all of which can be trained.

Every method in this book — from sand tables to simulations — contributes to developing these components. And the more fluently your soldiers can apply them, the more tactically capable your unit becomes.

But instruction isn't just about repetition. It's about creating understanding, not compliance. When your troops know *why* something works — not just *how* to do it — they're better prepared for situations you can't predict. They're more creative, more confident, and more adaptable. That's the mark of real leadership. The battlefield is chaotic. Your training shouldn't be.

We called this book *Dangerous Lessons* for a reason. Your responsibility as a leader and as an instructor is both serious and potentially dangerous. Like tactics. Like leadership itself. There are no final answers. There is only better thinking, better questions, and better practice.

And practice starts with you.

So go teach. Go lead. Build understanding. Shape your unit. Learn with them, grow with them — and make the next generation better than you were.

You owe your soldiers nothing less.

Good luck.

FURTHER READING

We offer you a short list below for your professional military education (PME). Some titles will expand your thinking, some will spark tactical ideas. The list is ordered by what you should read first. A word of warning: PME is a slow process and so it should be. Take your time and absorb the lessons.

The Art of War - Sun Tzu (Griffith translation)

Arguably the oldest of the worthwhile texts on war. This book is a compilation of aphorisms and stories intended to give the reader an insight into the nature of war. It is NOT a "how to" text. It is an introduction into a Taoist philosophy and should be read in that light.

The Defence of Duffer's Drift - E.D. Swinton

Swinton, the father of the tank, writes an insightful and often humorous account of the young subaltern, LT Backsight-Forethought, who in a series of six dreams, repeatedly kills everyone in his command until he ultimately learns enough about tactics to successfully defend the fictional Duffer's Drift during the Boar War. Although Captain (later Major General) Swinton published it as a fictional tale, his aim was to teach tactical lessons and generate discussion and debate among subalterns.

The Maneuver Warfare Handbook - William Lind

Very few pamphlets written by non-military people can claim to have changed warfare. This one did just that. This syncretic compilation of German warfare and Sun Tzu upset all NATO doctrine. Maneuver Warfare, although not yet fully understood was created by Lind with the help of his USMC coauthor.

On Infantry - John English & Bruce Gudmundsson

This 1994 revised edition tells the story of infantry in the 20th century and offers interesting insights on the role that infantry. There is wide use of foreign sources generally unavailable to the English-speaking reader. Combat motivation, the role of squads and fire teams, the role of infantry in the Blitzkrieg, and more is discussed and analyzed in this revised edition.

The Art of War in the Western World - Archer Jones

Consider this book as one of your go-to references for general military knowledge. Jones nicely combines three major components of war (tactics, strategy and logistics) to explain the last 2,500 years of military history, from phalanxes in ancient Greece through to the Thirty Years' War that shaped modern Europe.

The Face of Battle - John Keegan

Keegan, for many years a history professor at Sandhurst, does a good job of trying to put the reader in the front lines of three of history's greatest battles. Groundbreaking when published, it looked at the mechanics of battle at Agincourt, Waterloo and the Somme. Note that it is written from a British perspective.

Battle Leadership - Adolf von Schell

Von Schell was a Prussian infantry officer and after World War I he wrote of his experiences. This book offers many lessons for the junior leader including handling different personalities under stress, preparing troops psychologically for combat and more. Lots of lessons on leadership and tactics.

Infantry Attacks! - Erwin Rommel

A timeless classic that was overlooked when written. Many of the tactical lessons may seem self-evident, but you must keep in mind that it was all new at the time. The real value of this text is to gain appreciation for how important it is to be innovative and mentally flexible. Patton studied it before facing Rommel in battle, and in the movie famously (and apocryphally) shouts "You magnificent bastard, I read your book!"

Fighting Power - Martin Van Creveld

Van Creveld dissects the US and German Armies and their relative performance during World War II. Although some of van Creveld's conclusions have been controversial, his analysis is solid and worthy of study. This is an excellent general reference on warfare.

Dispatches - Michael Herr

John le Carre claimed that this was the best book he had ever read on men and war in our time. It is Herr's memoir of riding with the 1st Air Cav in Vietnam, from 1967-69. It is a "non-fiction novel," beautifully written but at times disturbing.

On Killing - Dave Grossman

This book expands on the famous work of SLA Marshall, Men Against Fire. It is a book that will give a good leader cause for pause as the author discusses the sometimes extraordinary and unexpected psychological and personal consequences of killing.

The Challenge of Command - Roger Nye

This book was specifically written to assist officers in their preparation for taking command. Although it is written by an American for other Americans, it is full of useful thoughts on the commander as leader, trainer, tactician, mentor and so on.

The Maneuver Warfare: An Anthology - Richard Hooker, ed.

A terrific compilation of essays arguing the many dimensions of Maneuver Warfare, both in theory and in practice. Many of the leading protagonists are here, including Robert Leonhard, William Lind, John Antal and Michael Wyly.

The Staff Ride: A Planning Guide - Andrew B. Godefroy

A concise guide to the planning and conduct of staff ride, which can be read in conjunction with a planned series of specific staff ride handbooks for junior leaders.

Websites of Note

Line of Sight
Although publicly available. it acts as the professional development hub for the Canadian Army and provides a large number of resources about tactics and doctrine. https://www.canada.ca/en/army/services/line-sight.html

The Cove
An Australian Army website that is their hub for military professional development. https://cove.army.gov.au

The Army Leader
The British Army's unofficial home for military professional development tools. https://thearmyleader.co.uk/

Fight Club
An international organization dedicated to wargaming and simulation as a means to improve military training. www.fightclubinternational.org

2nd Battalion, 5th Marines
A personal website that contains a vast number of resources related to teaching tactics and conducting operations. https://2ndbn5thmar.com

ANNEX A: SYSTEMS APPROACH TO TRAINING

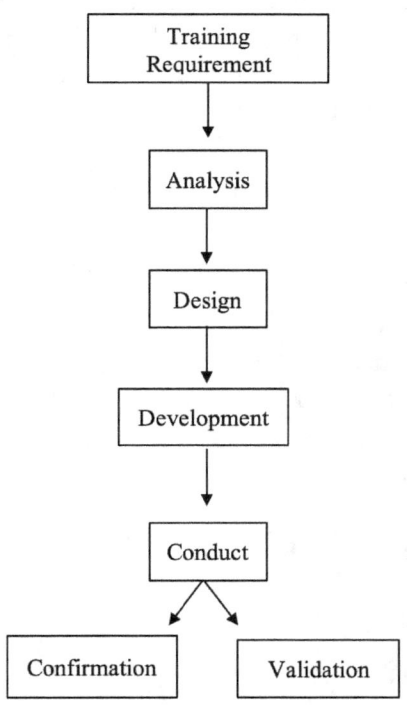

At left is a flow chart that describes how training is conceived, created and then conducted. It is the most common methodology used in NATO militaries, but is not the only way to do things. We recommend it because it works well and is easily understood. It always begins with understanding what the requirement at hand is. Once that is understood, a full analysis of that requirement is performed. That leads the trainer to designing exactly what type of training (among the many we have offered in this book) best suites the requirement. The next step is to flesh out the design with all of the necessary details. (Consider using the checklist at Annex B).

Once all of the preparatory work has been completed, the training can be conducted, again, using one or more of the methodologies we have offered above. The final step is bifurcated because it depends on whether your training was designed to confirm that skills were learned or whether you were validating an existing set of skills. Either way, if the answer is yes, training is complete. If the answer is no, re-training is required.

Don't forget the AAR!

ANNEX B: TRAINING DESIGN CHECKLIST

Checklists are valuable tools to help you ensure that you haven't inadvertently forgotten something. We all make mistakes, and a checklist can help you avoid them. The basic checklist offered below is generic enough that it should act as a good point of departure to help you begin the process of deciding, designing, creating and conducting whatever training you do.

- Who is conducting the training?
- What is the Training Aim?
- What is the Scope of the training?
- What are the Training Objectives?
- When will the training be conducted?
- Who is the Primary Training Audience (PTA)?
- Is there a Secondary Training Audience (STA)?
- What friendly ORBAT will be used?
- What enemy ORBAT will be used?
- What map (or physical location) will be used?
- Will an existing training scenario be followed?
- Will a new training scenario be required?
- What administrative support will be needed?

ANNEX C: AFTER ACTION REVIEWS

There are many ways to conduct an After Action Review, but most manuals are aimed at higher levels than we are discussing here. We thereby would like to offer a simplified template for you to consider, as well as a couple of "rules".

Remember: as the commander, conducting an AAR is *your* responsibility. Why? Because the training was *your* responsibility.

Use an 8-Step Approach

1. Introduction
2. Enemy Threat
3. What was Planned
4. What Really Happened
5. Open Discussion
6. Self-Assessment
7. Optional Issues
8. Summary

Rules for all Participants

1. Leave your ego at the door!
2. Everyone needs to speak from their perspective.
3. If it went well – feel free to point out the good efforts of others.
4. If it went not so well – concentrate on your personal experiences.
5. Disagreement is not disloyalty.
6. Facilitate, DO NOT dictate.
7. Be constructive: identify a problem; recommend a solution!
8. An AAR may be conducted at any point during a training activity.
9. If you train without an AAR, much of that training has been wasted.

What is a good AAR?

An effective After Action Review has the following characteristics:

- **Leader humility**. The commander sets the tone by admitting where they went wrong and showing a willingness to learn. This models the behaviour expected of everyone present and makes it clear that the AAR is about improving performance, not protecting egos.

- **Shared understanding**. By the end of the discussion, all participants should have the same grasp of the problems identified, the root causes, and the agreed solutions. Without this, individual interpretations may drift and the lessons will be lost in execution.

- **Actionable outcomes**. Issues raised must be within the power of the unit or organization to fix. If the solution depends entirely on outside agencies or unrealistic resources, it will be ignored. A good AAR produces changes that can be implemented immediately or built into the next training event.

- **Inclusive learning**. Valuable insights can come from any rank or role. Encouraging contributions from junior personnel often surfaces ground-level details that senior leaders missed, and it reinforces that everyone is responsible for improving performance.

Caution

It's a First Principle of learning that members of any organization will be more receptive to criticism and improvement when they identify it for themselves. So, whenever possible, learning points should be drawn from participants, not presented by the leader. Use leading questions to draw out points from the participants. This discovery technique encourages greater learning

ACKNOWLEDGEMENTS

The people who contributed to our knowledge about teaching tactics over the years are to numerous to recount fully, but it would be wrong to think that they were all officers or even our superiors when the lessons were taught. Teaching the tactics is the province of the professional soldier, but not of any particular rank, position or trade. To all of these people, were are grateful. We stand on the shoulders of giants.

The list of people who reviewed this manuscript and provided feedback is much more manageable. We greatly appreciate the time taken by Major Ed Farren (UK), Mr Nick Riggs (Bath Spa University), Major John Bosso, Major Matt Brennan, Major Morgan Oliviero and Lieutenant Alex McConnachie (Canada).

We particularly want to thank the folks at Fight Club International and Fight Club Canada, who have picked up the torch from their respective militaries and focus on using every available means to teach and practice tactics. Their professionalism and enthusiasm are incredibly admirable, and their efforts will undoubtedly serve to one day save lives on the battlefield.

ABOUT THE AUTHOR

Colonel Chuck Oliviero, PhD spent almost four decades in the Canadian Army as an Armoured Cavalry officer. He spent half of his career either as a student or as a teacher and the other half commanding troops. He is the author of three non-fiction books on war and strategy: *Praxis Tacticum*, *Strategia* and *Auftragstaktik*, as well as one novel, *The Cohort*.

ABOUT THE AUTHOR

Phil Halton is a Canadian Army veteran and the author of three novels, *This Shall be a House of Peace* (2019), *Every Arm Outstretched* (2020) and *Red Warning* (2024), as well as a history, *Blood Washing Blood: Afghanistan's Hundred-Year War* (2021). He has worked in conflict zones around the world as a security consultant, but is now a PhD candidate at the University of Gloucestershire.

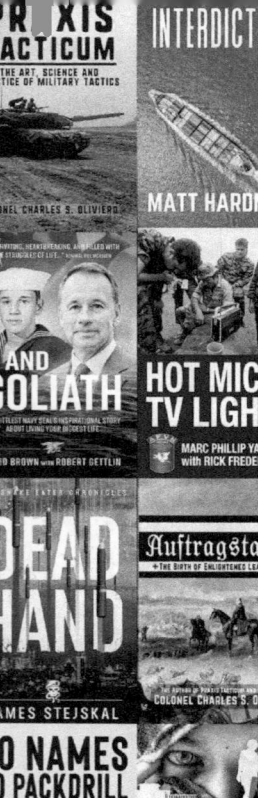

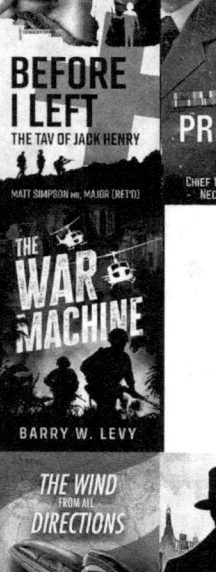

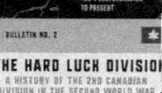
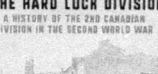

DOUBLE‡DAGGER
— www.doubledagger.ca —

ABOUT THE PUBLISHER

DOUBLE DAGGER BOOKS is Canada's only military-focused publisher. Conflict and warfare have shaped human history since before we began to record it. The earliest stories that we know of, passed on as oral tradition, speak of war, and more importantly, the essential elements of the human condition that are revealed under its pressure.

We are dedicated to publishing material that, while rooted in conflict, transcend the idea of "war" as merely a genre. Fiction, non-fiction, and stuff that defies categorization, we want to read it all.

Because if you want peace, study war.

NOTES

QUICK REFERENCE MATRIX

Method	Tactical Acumen	Tactical Awareness	Speed of Decision-Making	Tactical Agility
DOES THIS METHOD TEACH:				
1 - Sand Tables	✓	✓	✗	✓
2 - Field Problems	✓	✗	✓	✗
3 - Battlefield Studies	✓	✗	✗	✗
4 - TEWTS	✓	✗	✓	✗
5 - TDGs	✓	✗	✓	✗
6 - Tactical Discussion	✓	✓	✓	✗
7 - TTXs	✓	✓	✓	✗
8 - Stand Training	✓	✓	✓	✓
9 - Force-on-Force	✓	✓	✓	✓
10 - Live Fire	✗	✗	✗	✓
11 - Virtual Sim	✓	✓	✓	✗
12 - Constructive Sim	✓	✓	✓	✗
13 - Commercial Games	✓	✓	✓	✗
14 - LLM Sim	✓	✓	✓	✗